GENETICS
for Healthcare Professionals
A lifestage approach

GENETICS
for Healthcare Professionals
A lifestage approach

Heather Skirton
*Nurse Consultant in Clinical Genetics, Clinical Genetics Unit,
Taunton and Somerset NHS Trust, Taunton, UK
Lecturer, University of Wales College of Medicine, Cardiff, UK*

and

Christine Patch
*Genetic Nurse Counsellor, Southampton General Hospital,
Southampton, UK*

© BIOS Scientific Publishers Limited, 2002

First published 2002

A CIP catalogue record for this book is available from the British Library.

ISBN 1 85996 043 X

BIOS Scientific Publishers Ltd
9 Newtec Place, Magdalen Road, Oxford OX4 1RE, UK
Tel. +44 (0)1865 726286. Fax +44 (0)1865 246823
World Wide Web home page: http://www.bios.co.uk/

Important Note from the Publisher
The information contained within this book was obtained by BIOS Scientific Publishers Ltd from sources believed by us to be reliable. However, while every effort has been made to ensure its accuracy, no responsibility for loss or injury whatsoever occasioned to any person acting or refraining from action as a result of information contained herein can be accepted by the authors or publishers.

The reader should remember that medicine is a constantly evolving science and while the authors and publishers have ensured that all dosages, applications and practices are based on current indications, there may be specific practices which differ between communities. You should always follow the guidelines laid down by the manufacturers of specific products and the relevant authorities in the country in which you are practising.

Production Editor: Andrea Bosher
Typeset by Phoenix Photosetting, Chatham, Kent
Printed by Cromwell Press, Trowbridge, UK

Contents

Preface

The motivation to write this book arose directly out of many years of working in clinical practice in genetics. Part of our role has always been to share the knowledge we have with our clinical colleagues of all disciplines. It has been difficult to find educational material that presents genetic information in an accessible way that has relevance to practice. This book is an attempt to redress that balance.

Both of us were privileged to start working in clinical genetics in the late 1980's, at the beginning of the explosion of what has come to be known as the 'New Genetics'. At that time our counselling and information giving was based almost totally on assessing risk from the family tree, and combining that with information from family studies. Our theoretical knowledge was derived from textbooks, journals and from listening to our senior colleagues at work. There were no direct genetic tests for diagnosis or predictive purposes and the Human Genome Project was an idea, not a reality. Since that time most of the major single disease genes have been discovered and the Internet has facilitated the rapid transfer of current scientific knowledge. The technology used to identify genetic variation has developed exponentially and scientists, clinicians and policy makers are trying to predict and prepare for the next wave of genetic advances.

The fascination of medical genetics lies in the merging of exciting, fast developing science with the real world of people and their lives. We were privileged to develop our expertise by learning from the families we met in the clinical situation. This book would not have come to fruition without that experience and we would like to acknowledge all the individuals and families we have been involved with over the years. Our aim is to try and give you the reader enough theoretical knowledge to enable you to access the information you need. We hope that in placing the theoretical knowledge in the context of hypothetical family stories we are able to show that genetic science and technology is relevant to your practice.

Genetic technology in health holds the potential for benefit but also harm. Health care practitioners have a responsibility to identify the knowledge they need for practice and fill those knowledge gaps. We hope that by sharing our experience with you, you are able to feel confident in considering the implications of genetic science for your practice and that you are encouraged to develop your knowledge further whatever your clinical background or speciality.

Christine Patch and Heather Skirton

Acknowledgments

Our thanks to: Sally Boxall, Lauren Kerzin-Storrar, Hazel Ross and Angus Clarke for reviewing the draft and for their many helpful suggestions; Dr Peter Lunt and Dr Nick Dennis, our first mentors in clinical genetics; our colleague, Chris Barnes, for special encouragement and support; Katherine Farthing for help with the website references; and our families for their patience.

This book is dedicated to Chris and Stephen, the wind beneath our wings.

About the authors

Heather Skirton – has extensive experience in pediatric nursing and midwifery, became a nurse genetic counsellor in 1989, and has since completed her Master's degree in Health Care and obtained a PhD. The first Nurse Consultant in Genetics in the UK, heather divides her time between clinical practice and teaching on a Master's course in genetic counselling. Heather is also a Course Director at the European School of Genetics and Chair of the Global Concerns Committee of ISONG.

Christine Patch – became a genetic nurse counsellor in 1988 and has a Master's degree in Health Professional Education. Christine is currently studying for her PhD and is developing an evidence base for genetic healthcare. Christine was the first General Secretary of the BSHG and is actively involved in raising the profile of genetic nurses and counsellors throughout Europe.

Notes to readers on the format of this book

- Key practice points are included in boxes in the text, to emphasize important clinical concepts.
- An extensive glossary is included at the end of the book. Where a word is encountered for the first time in the text, it is marked in bold to indicate that it is included in the glossary.
- Appendix I is a website resource. All websites cited in the text are included in the appendix, as are many others that may be useful to you.
- Worked answers to the problems set at the end of each chapter are in Appendix V.
- Although all examples are based on our work with families over many years, the names and identifying details have been altered so that no living or deceased person is represented. In most cases, each scenario represents an integration of the experience of several different families.

Setting the scene

1. Introduction

This book has been written for all health professionals who have an interest in genetics. Increasingly, those working in primary and secondary care find they are required not only to understand the basics of genetics, but also to explain the principles to clients and to support them through genetic testing or screening. This is especially true of nurses, midwives and health visitors.

In the past, genetics was a subject confined to a number of rare conditions, and the number of families who required these explanations was relatively small. However, with the advent of the '**new genetics**', genetic principles of testing and treatment are being applied to a larger section of the population, and the need for genetics knowledge is therefore imperative for a wider range of health professionals [1].

It is now accepted that genetics has changed the face of medicine by introducing a 'new taxonomy of disease' [2]. Diagnosis and treatment of common diseases such as cancer, heart disease and diabetes will be based upon knowledge of the genetic alterations that influenced the disease and the patient's response to treatment. Although gene therapy is still not in clinical use, **gene** testing is now being used to identify clients at high risk of bowel [3] and breast cancer [4], so clinical screening can be targeted. The widespread use of antenatal **maternal serum screening** [5] and fetal anomaly scanning [6] means that every midwife needs to be aware of the use of genetic tests and the benefits and limitations of such tests. The changing health scene reflects the increasing influence of genetics, but it is apparent that the education of health professionals has not kept pace with these changes.

It is hoped that this book will enable those working in health care to update their knowledge and apply it usefully in patient care. This book is primarily aimed at those who are starting work in **clinical genetics**, those who work in closely related fields such as midwifery, health visiting, oncology and neurology, and those who have an interest in genetics and wish to deepen their own knowledge. The information is organized in chapters relating to particular life stages. Throughout the book, case examples are used to demonstrate the clinical application of the topic under discussion and the six families introduced in this first chapter are used to explore clinical situations. An extensive glossary is included at the end of the book, with definitions of any terms that may be unfamiliar, and all websites cited are included in Appendix I. In this chapter, some important concepts related to genetic health care are discussed to set the scene.

2. Defining genetic counseling

One definition of genetic counseling that is generally accepted was written by the American Society for Human Genetics in 1975 [7]. Although a quarter of a century has passed since it was first written, it still accurately reflects the extent of services provided under the title 'genetic counseling.' It states that genetic counseling is:

> "a communication process which deals with human problems associated with the occurrence, or the risk of occurrence, of a genetic disorder in a family. This process involves an attempt by one or more appropriately trained persons to help the individual or family to (1) comprehend the medical facts, including the diagnosis, probable course of the disorder, and the available management; (2) appreciate the way heredity contributes to the disorder, and the risk of recurrence in specified relatives; (3) understand the alternatives for dealing with the risk of recurrence; (4) choose the course of action which seems to them appropriate in view of their risk, their family goals and their ethical and religious standards, and to act in accordance with that decision; and (5) make the best possible adjustment to the disorder in an affected family member and/or the risk of recurrence of that disorder."

3. Genetic services

Genetic services are provided in many countries, including the UK, USA, Canada, Australia and The Netherlands, by a variety of health professionals who often work together in multidisciplinary teams. In the UK, specialist genetic services are provided by staff working in regional genetic centers. The team includes medical geneticists and **genetic counselors**. Although the majority of genetic counselors have a background in nursing or midwifery, increasingly they may enter the profession by undertaking a Master's degree in genetic counseling [8]. Most families are referred to the genetic service by either the family doctor or a specialist (e.g. pediatrician, neurologist or obstetrician) caring for the **affected** or at risk person. In most centers, however, referrals will be accepted from any health professional.

The types of families referred for genetic services fall into four main groups:

1. Families who are concerned about a known or suspected genetic condition such as cystic fibrosis, Huntington's disease or muscular dystrophy.

2. Where there is concern about a child in the family who has learning **delay** and/or **dysmorphic features**.

3. After the diagnosis of fetal abnormality or fetal loss during a pregnancy or in the neonatal period, or recurrent **spontaneous abortion**.

4. Families in which there is a strong history of cancer.

The following six referrals are typical of those received in any genetics department, and these families are used throughout the book to illustrate the principles discussed. Other family examples are given where relevant.

Harding Family

Dear Doctor

Re: Sarah Harding age 22

I would be grateful if you could see Sarah and offer her some advice regarding her family history of Huntington's disease.

Sadly, Sarah's mother suffers from this condition and Sarah is now wondering whether she could be tested, especially as she is expecting her first baby. We would welcome your expert advice.

Yours

Community Midwife

Collins Family

Dear Doctor

Re: Carol Collins, age 33 years

Adam Collins, age 11 years

Carol has been in the surgery today with her son Adam aged 11. In the course of the consultation for another reason she mentioned that she'd been told that Adam would need a bowel check from about now. Carol herself had a total colectomy some years ago and still has regular check-ups at the hospital. She tells me that this is a genetic condition and when she had her operation she was told the children should be screened when they reached the age of 10. I would be grateful if you would see and advise.

GP

Inskipp Family

Telephone referral received from neonatal pediatrician, notes taken by the secretary are:

Baby Inskipp

Born today, small for dates

Cleft lip and palate

Microcephaly

Unusual ears

Please give an opinion

Although the service for individual families varies according to need and the condition concerned, in general the following aspects of care will be offered by the genetic service.

Diagnosis. Although this seems an obvious starting point, in many situations the correct diagnosis of an affected member may not be known because of limited medical knowledge at the time the person became symptomatic, or their reluctance to pursue the diagnosis. In order to treat an affected person, the exact mode of inheritance may not be relevant, but to advise family members about the risk to themselves or their offspring, it is a critical piece of information. Let's think about Melanie, a young woman who presents asking about the risk of muscular dystrophy. Her brother aged 20 years is affected, and she is worried about her future children. She says her brother has muscular dystrophy, but this could be one of many types, with different patterns of inheritance. It could be:

i. Duchenne muscular dystrophy, an **X-linked** condition. Women can be **carriers**, and if she were a carrier then each male child of hers would have a 50% risk of having the condition.

ii. Myotonic dystrophy, a **dominant** condition. She could have very mild signs and if she were affected both male and female children would have a 50% risk of inheriting muscular dystrophy. Her children might be at risk of a very severe congenital type of myotonic dystrophy.

iii. Limb girdle muscular dystrophy. This is an **autosomal recessive** condition and, even if she were affected, her children would be at low risk because the chances that her partner also carried the condition would be very low.

Finding the correct genetic basis of the muscular dystrophy is necessary to give Melanie the correct information about the risks to her future children and the potential for prenatal testing. Similarly, a family history of cancer must be confirmed to enable correct risk estimates and accurate advice about screening to be given. This is covered in detail in *Chapter 9*.

Information giving includes:
- mode of inheritance;
- natural history of the condition;
- signs and symptoms;
- prognosis.

Evaluation of risk and discussion and explanation of that risk to each interested family member.

Presentation of appropriate options, including:
- preventive/lifestyle measures (e.g. use of folic acid pre-conceptually to reduce risk of **neural tube defects**);
- testing, including diagnostic, presymptomatic, predictive and prenatal;
- clinical surveillance such as mammography or renal scans.

Ongoing contact and support. Genetic counseling does not generally include treatment or management of the genetic condition, which is generally organized by the relevant speciality service. However, it is true that some genetic conditions are so rare that affected individuals may attend a specific genetic clinic to be seen by a genetic specialist who has particular expertise in the management of that condition, for example, neurofibromatosis type II.

4. Types of genetic testing

4.1 What is a genetic test?

There has been much discussion about the definition of a genetic test. Although any test involving **DNA** or **chromosome** studies may be considered a genetic test, some such tests do not actually provide information about the person's genetic identity. A good example is the use of chromosome studies in the diagnosis of leukemia, in which the Philadelphia chromosome is an indicator of the presence of the disease rather than an inherited characteristic. However, 'routine' tests may reveal information that has relevance for both the individual and other members of the family. This might occur if a full blood count results in a diagnosis of thalassemia or sickle cell anemia. A helpful working definition of a genetic test is, therefore, any test the result of which would prompt the health professional concerned to alert other family members of potential risk. In reality, the health professional would normally ask the original patient to inform the family, rather than contacting the family personally, but it is the principle of broader contact that is important rather than the practicalities.

4.2 Diagnostic test

Diagnostic genetic tests are those performed where there is already a clinical suspicion of disease, and the test is performed to confirm or identify the diagnosis.

4.3 Predictive test

Usually performed in a family in which there is a known genetic condition, predictive tests are carried out prior to the onset of any symptoms, and confirm the presence of a gene **mutation** that will eventually result in disease. Because these conditions are usually those that occur in adulthood, predictive tests are rarely carried out on minors. Free choice in opting to be tested, and psychological preparation for the result are considered important [9] (see *Chapter 9*).

4.4 Presymptomatic test

Presymptomatic testing differs from predictive testing in that it is carried out for conditions in which clinical problems associated with the gene mutation do not necessarily occur. For example, a woman may have presymptomatic testing for a breast cancer gene mutation, however, she will not always develop breast cancer even if she has the mutation (see *Chapter 9*).

4.5 Prenatal test

This applies to any genetic test performed to provide information about the status of the fetus, and includes results of ultrasound scanning, chromosome studies performed with material obtained during amniocentesis, **chorionic villus biopsy** or fetal blood sampling (see *Chapter 6*).

Genetic testing differs from **genetic screening**, a term used for testing of a population in order to detect those at high risk. Genetic tests are performed for an individual at significant risk because of family history or because they are symptomatic, for example, testing the fetus of a woman who is a known carrier of Duchenne muscular dystrophy. Measuring serum creatine kinase levels of all newborn boys to detect raised levels (that may indicate the boy has a form of muscular dystrophy) is an example of genetic screening.

5. Ethical practice

The ethical principles underlying 'value' ethics have long been the mainstay of medical ethics [10] and these underpin the work of the genetic counselor. The main principles are:

- beneficence (the duty to do good);
- nonmaleficence (the duty to prevent harm);
- justice (equity of care);
- autonomy (the belief that a person is able to act in their own best interests).

However, when working with a number of individuals within a family, these principles may not provide straightforward guidelines or solutions.

ETHICAL DILEMMA   **ROY**

Roy is found to have a balanced chromosome **translocation** after his wife has three spontaneous abortions. If one parent carries a translocation, there is an increased risk in every pregnancy of either having a child with physical and mental disabilities or spontaneous abortion.

Roy has two sisters, who may also carry the translocation. He refuses to tell them of his carrier status and alert them to the risks to their own future children.

Respecting Roy's autonomy by maintaining his confidentiality may conflict with the rights of his sisters to prevent the birth of a child with a serious disability.

ETHICAL DILEMMA   **JOAN**

Joan is 45 years old and has a 50% chance of inheriting Huntington's disease (HD) from her mother. Her son Lindsay wants to be tested for the gene mutation that causes HD. He is engaged and wants to know his own status before deciding to have a family.

A positive result for Lindsay would inform Joan that she has also inherited the mutation. Joan dreads having HD and feels she would not cope if she knew that was awaiting her.

Benefiting Lindsay could result in harm to Joan.

In practical terms, these difficult situations are not usually resolved quickly. The genetics team will spend time with the family discussing the options for action, and potential outcomes. For example, the counselor might ask Lindsay how he might feel if his mother became depressed after hearing the result, whereas Joan might be asked to consider how she might feel if Lindsay and his wife decided they could not risk having a family without having a test. Often over time the family will resolve the conflict by finding a compromise. When Lindsay realized how his mother felt he agreed to wait until she was 50 before being tested. As Joan's mother was affected by the age of 40 years, if Joan had inherited HD she was likely to be symptomatic by that time. Joan agreed to this plan, as she did not wish to deny Lindsay the chance to find out and felt she had time to adjust to prepare for knowing her status.

If the family is unable to negotiate a solution, then the genetics team will discuss a course of action that causes least harm, usually seeking opinions from other experienced team members and colleagues in other regions or specialities. The opinion of a medical ethicist or lawyer might be sought. In Roy's case, he was unable to change his position. The genetics team felt the potential harm to his sisters outweighed his request for confidentiality, and informed the sisters' doctor of the advisability of them being offered a chromosome test. Roy was informed of this action, and his confidentiality was preserved as far as possible by the wording of the information disclosed.

6. Genetic testing of children

Genetic testing of children is another potentially controversial area of practice. Children are generally not tested unless the result will in some way benefit the health of the child before they reach adulthood [11]. If the test result will only be of significance in adulthood, then the test is usually deferred until the child is able to give informed consent. Informed consent implies that the person consenting has understood the reason for the test and the implications of the result.

In some cases, genetic testing in childhood will be of benefit to the child. There are often benefits from making a diagnosis in a child with a genetic

---

CASE EXAMPLE **SAMANTHA**

A child called Samantha was referred to the genetic service because of her learning problems at school. She was falling behind her classmates in Year Two. A chromosome study showed she had a **microdeletion** of chromosome 22, which can cause learning difficulties, speech delay and heart abnormalities. Samantha had an echocardiogram that showed she had a mild heart defect, and she was treated by the pediatric cardiologist. She was also referred for speech therapy, and because of her diagnosis was granted additional help at school.

syndrome, as there may be problems associated with the syndrome that would warrant extra surveillance or treatment.

In certain cases, a young person who has known of his/her genetic risk for many years will request a carrier test before they are 18 years of age. Provided the request comes from the young person, and they are able to give informed consent, the request for testing will probably be seriously considered.

In some cases, the child may be at risk of a condition for which invasive screening is recommended. Genetic testing for the child may clarify the need for such screening.

Collins Family

*Robert and Gemma are the two children of Michael, the brother of Carol Collins, who has familial adenomatous polyposis (FAP). This is a condition in which multiple **polyps** grow in the colon, predisposing the affected person to colorectal cancer. Screening by **colonoscopy** is usually arranged for such children annually from about the age of 12 years. A genetic test that clarifies whether the children have inherited the gene mutation from their father would either confirm the need for invasive screening, or remove the necessity for the screening. In this case, genetic testing is clearly in the interests of the children. However, the test would usually be undertaken as late as possible before screening starts, to give the children the opportunity to discuss it fully and give their consent. In this family, Gemma was tested at 11 years and was found to have the mutation. She had screening annually and had a colectomy at 17 years of age. Robert was tested the following year, when he was 10 years, at his request. He had not inherited the mutation and screening was not necessary for him.*

Guidelines on the testing of children have been written by the Clinical Genetic Society (UK) and copies are available from them.

7. Genetics and insurance

There has been considerable debate in the literature and media about the potential for discrimination, both in employment and insurance, if predictive genetic testing is carried out. The concern is that if an individual has a genetic test that gives a risk of specific illnesses developing that individual will be unable to purchase insurance [12]. However, it should be borne in mind that if an individual discloses a family history of a genetic condition, this will itself have an impact on obtaining insurance and the premiums paid. It is not only genetic testing that affects the situation, but family history. Because most insurance application forms include questions on family history, it is not possible to avoid providing this information, and to knowingly withhold it may make the insurance invalid.

It should be remembered that the potential impact of genetic discrimination will be very different according to public policy in individual countries. For example, in the USA where much of the concern is expressed, many of the population depend on private insurance for their healthcare provision, and consequently an individual alteration in risk will have implications for their insurability. In the UK where health care is funded through taxation (at the present time) and is free to all at the point of service, the issues of genetic testing in relation to healthcare provision are different. This may become more relevant in the UK if genetic testing for predisposition to elderly-onset diseases such as dementia becomes possible, as funding for nursing home care for the elderly is constantly under review. In the UK discussions between the government, the insurance industry, the clinical genetics community, consumers and other stakeholders have led to a five-year moratorium (from October 2001) on the use of genetic tests in assessing applications for life insurance, critical illness, long-term care and income protection policies up to specific financial limits. In addition an advisory body, the Genetics and Insurance Committee, assesses the actuarial validity of any proposed genetic test, and advises on its potential usefulness. It is essential that dialog continues between the involved parties in order to insure that both the potential and limitations of genetic advances are fully discussed. The Public Health Genetics Unit website (see *Appendix I*) provides background and relevant reading for this topic.

8. Impact of the genetic condition on families

It is extremely difficult to generalize about the impact of a genetic condition on the family, as the conditions, circumstances and families differ so greatly. Perhaps the best way to approach each family is to bear in mind that one cannot predict the impact. What may be devastating for one family may be seen positively by another.

For most families, however, the diagnosis of a genetic condition is accompanied by a feeling of loss. This may be related to:
- loss of family member due to death;
- loss of health;
- impairment or disability;
- loss of a normal future or the expectation of a normal future;
- loss of independence;
- loss of security.

Because of the familial nature of genetic conditions, these losses often extend to the wider family, and the family support systems are stretched. Many family members report that they feel guilty, this might be because they have 'passed it on,' however unknowingly or even because they have escaped the

CASE EXAMPLE   JOANNE

Joanne is a genetic counselor. On the same day she was asked to see two different couples, both of whom had recently terminated a pregnancy.

The first couple she saw were devastated, it had been their first pregnancy and they 'had not dreamt anything could go wrong'. When the baby was diagnosed with severe **spina bifida** and **hydrocephalus** at 19 weeks' gestation, they had decided to terminate the pregnancy but felt guilty, as though they had abandoned her. Joanne spent an hour with them, although little information was discussed at the time because they were so distressed. She saw this couple on three further occasions.

She went on to visit another couple who had also terminated a recent pregnancy, after a scan showed a serious heart defect. This couple were farmers who felt that an abnormal baby had a poor chance of survival, and that termination of pregnancy had spared their child additional suffering. They were already planning another pregnancy and wanted to insure the fetus was scanned at an early stage.

disease, whereas other relatives have not. Guilt remains a part of many parents' lives, and the loss of the freedom to have a family without undue worry is an aspect that features strongly in some family stories [13].

In any experience of loss, mourning has a part in helping the family adjust to the loss and resume life within their altered circumstances. The pace at which family members achieve this will often differ, and this may cause additional family tensions. In our experience, it is often helpful to confirm that a loss has occurred and that feelings of grief are natural. Brock [14] has written of the therapeutic power of telling your story. For professionals working with those at risk of a genetic condition, allowing time to listen to the story can be genuinely beneficial to the family.

9. Nondirectiveness in genetic health care

The aim of genetic counseling is broadly to enable individuals and families at genetic risk to live as normal a life as possible. However, the focus is very much on the family's wishes. The onus is therefore upon the genetic counselor to present the available options in a nonjudgmental way and to facilitate the client to make the best choice for them, in their unique circumstances. Much has been written about the need for **nondirectiveness** in genetic counseling [15] and this may be, at least in part, a reaction to the eugenic approaches adopted in some countries during certain periods of history [16].

Nondirectiveness by the counselor is not synonymous with remaining 'uninvolved' in the client's process of decision-making. Ideally, the counselor is skilled enough to enable the client to explore the options and support them in their choice, while remaining nonjudgmental. It is recognized this is a challenging task, particularly if the personal values and beliefs of the counselor conflict with those of the client. For this reason, training in counseling skills that includes work on developing personal awareness is a necessary part of the preparation and ongoing education of genetic counselors [17]. Counseling and clinical supervision is also an ongoing requirement for work in this field.

10. Documentation

As in all healthcare settings, thorough documentation of the care given and the clinical management of the case is essential. The notes made by the practitioner may have an impact not only on the current client(s), but on family members in the future. Where possible, confirmation of a diagnosis is obtained before a risk genetic assessment is made and options presented to the client. This may involve tracing old medical notes, death certificates, post-mortem reports, pathology or X-ray reports, or genetic test results. If the relative concerned is living, written consent should be sought to examine the medical notes. A detailed account of the evidence obtained and the information given to the client can (and probably will) be used to offer counseling to other family members in the future. A summary letter to clients is a useful means of recording what was said, not only for the client but for the medical record.

11. Conclusion

In this initial chapter, we have introduced some of the relevant concepts for health professionals who provide care for any individual or family with concerns about a genetic condition. In the following chapters, we focus on families at different times in the life cycle, and highlight the genetic issues that are relevant for each.

TEST YOURSELF

Q1. What are the components of a genetic counseling interaction?

References

1. Collins FS, McKusick VA (2001) Implications of the human genome project for medical science. *JAMA* **285** (5): 540–544.
2. Department of Health (1995) *The Genetics of Common Diseases. A Second Report to the NHS Central Research and Development Committee on the New Genetics.* London: Department of Health.
3. Vasen HF, Mecklin JP, Khan PM, Lynch HT (1991) The International Collaborative Group on Hereditary Non-Polyposis Colorectal Cancer (ICG-HNPCC). *Dis Colon Rectum* **34** (5): 424–425.

4. Claus EB, Risch N, Thompson WD (1991) Genetic analysis of breast cancer in the cancer and steroid hormone study. *Am J Hum Genet* **48** (2): 232–242.

5. Wald NJ, Kennard A, McGuire A (1998) Antenatal testing for Down's syndrome. *Health Technol Assess* 2(1).

6. De Vigan C, Baena N, Cariati E, Clementi M, Stoll C (2001) Contribution of ultrasonographic examination to the prenatal detection of chromosomal abnormalities in 19 centres across Europe. *Ann Genet* **44** (4): 209–217.

7. Ad Hoc Committee on Genetic Counseling American Society for Human Genetics (1975) Genetic counseling. *Am J Hum Genet* **27**: 240–242.

8. Skirton H, Barnes C, Curtis G, Walford-Moore J (1997) The role and practice of the genetic nurse: report of the AGNC Working Party. *J Med Genet* **34** (2): 141–147.

9. IHA website (Accessed 2001).
http://www.huntington-assoc.com/ihanews.htm

10. Campbell A, Charlesworth M, Gillett G, Jones G (1997). *Medical Ethics*. Oxford: Oxford University Press.

11. Clinical Genetics Society (1994) *The Genetic Testing of Children*. Clinical Genetics Society, British Society of Human Genetics Office, Birmingham, UK.

12. Chadwick R, ten Have H, Hoedemaekers R *et al.* (2001) Euroscreen 2: towards community policy on insurance, commercialization and public awareness. *J Med Philos* **26** (3): 263–272.

13. Faulkner CL, Kingston HM (1998) Knowledge, views, and experience of 25 women with myotonic dystrophy. *J Med Genet* **35** (12): 1020–1025.

14. Brock SC (1995) Narrative and medical genetics: on ethics and therapeutics. *Qual Health Res* **5** (2): 150–168.

15. Kessler S (1997) Psychological aspects of genetic counseling. XI. Nondirectiveness revisited. *Am J Med Genet* **72** (2): 164–171.

16. Kevles DJ (1995) *In the Name of Eugenics*, 2nd edn. New York: Alfred A Knopf/Harvard University Press.

17. Skirton H, Barnes C, Guilbert P *et al.* (1998) Recommendations for education and training of genetic nurses and counselors in the United Kingdom. *J Med Genet* **35** (5): 410–412.

Further reading

Campbell A, Charlesworth M, Gillett G, Jones G (1997) *Medical Ethics*. Oxford: Oxford University Press. (A general text on medical ethics, with some discussion on ethics and genetics health care.)

Harper PS, Clarke AJ (1997) *Genetics Society and Clinical Practice*. Oxford: BIOS. (Discussion of a broad range of topics related to clinical genetics and social and ethical issues.)

Lea DH, Jenkins J, Francomasco C (1998) *Genetics in Clinical Practice: New Directions in Nursing and Healthcare*. Sudbury: Jones and Bartlett Publishers. (General textbook written by nurses covering many aspects of healthcare genetics.)

2 The family history

1. Introduction

Although there are a great number of genetic tests conducted in a laboratory, the single most important tool and test in clinical genetics is the family history. Taking an accurate family history gives the genetic counselor a wealth of information about the likelihood of a genetic condition in the family, the probable inheritance pattern, and **recurrence risks** for family members.

The ability to take a family history should be an essential skill for most health professionals reading this book! Following the guidelines given here, you ought to be able to attain a reasonable level of competence, but practice will help you develop more effective ways of obtaining the information needed from the client.

It is important to remember the ethical aspects of taking the family tree. Some of the information may be highly confidential, and should not be shared with other family members unless consent is expressly given.

2. Guidelines for drawing the family tree

1. There is a list of recognized symbols used in **pedigree** drawing, to enable others to correctly interpret the information. These symbols are illustrated in *Figure 1*.

2. Start with the client and his/her immediate family.

3. Build the tree in a structured way, asking first about one side of the family, then moving to the other side.

4. Remember to ask about pregnancy losses, you can do this by asking 'Did you lose any babies?'.

5. Clients may not remember to tell you about relatives who have died. One way to put this is 'and how many brothers and sisters *did* your mother have?', rather than 'how many brothers and sisters *does* your mother have?'.

6. Clients may include information about relatives who are not related biologically. It is important to acknowledge the social structure of the family as well as the genetic structure, for this reason we include those relatives, making a note on the pedigree.

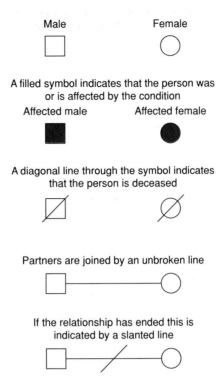

Male   Female

A filled symbol indicates that the person was or is affected by the condition

Affected male   Affected female

A diagonal line through the symbol indicates that the person is deceased

Partners are joined by an unbroken line

If the relationship has ended this is indicated by a slanted line

Figure 1. There is a generally accepted set of symbols that re used to convey information about the family structure via a pedigree. The word pedigree derives from the French for foot of a crane, as the original pedigrees resembled the structure of a bird's foot.

7. A note 'adopted' on the pedigree does not make it clear to someone else whether the relative has been adopted into the family or was given up for adoption, so insure this is recorded.

8. Ask about **consanguinity** in the family, as this may have some bearing on recessively inherited conditions. The question may be sensitive, but some ways of phrasing it gently are 'Do you have any grandparents in common?' or 'Were you related before marriage?'.

9. A client may offer very sensitive information about previous pregnancies or paternity issues. Use your discretion about whether these are relevant for documentation purposes. Remember that you may be using the pedigree when counseling the family in future, and need some way of reminding yourself about these issues.

10. When counseling different branches of the same family, it is helpful to draw a new tree, to ascertain what information is known by those family members. It is not uncommon to have completely different levels of information held by different family members. Discussion can then be restricted to the information held by that particular branch, without breaching confidentiality of other members of the family.

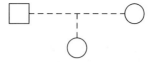

If a child's parents have never been a couple, the line between them is interrupted

Consanguinity is marked by a double line joining the two partners

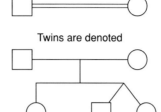

Twins are denoted

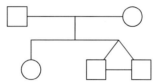

If they are monozygotic (identical) a horizontal line is drawn between them

A child who is stillborn is marked by a small circle, as is an abortion (whether spontaneous or induced). The gestation of the pregnancy is usually indicated if possible

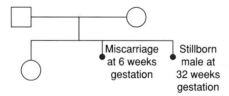

Miscarriage at 6 weeks gestation

Stillborn male at 32 weeks gestation

Figure 1. continued.

11. When you have finished drawing the tree, always ask, 'Is there anything you think I should know that we haven't mentioned?'.

More information and hints on drawing the pedigree are available on the Clinical Genetics Society website.

3. Drawing the tree – a practical example

David and Sally Inskipp have been referred to the genetics clinic with their daughter Maria. Maria is only 2 weeks old when the genetic counselor (C) visits the family to take a family history. Here is a transcript of that session.

C: Can we start by drawing a family tree?
Sally: I'm not sure we can tell you much, we're not a very close family, and David is adopted so we don't know anything about his side.

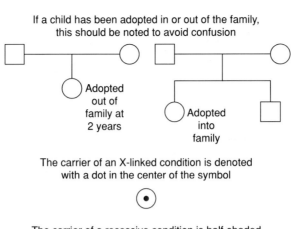

If a child has been adopted in or out of the family,
this should be noted to avoid confusion

Adopted
out of
family at
2 years

Adopted
into
family

The carrier of an X-linked condition is denoted
with a dot in the center of the symbol

The carrier of a recessive condition is half-shaded

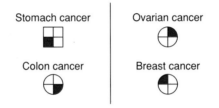

If a condition may be manifested in a number of forms,
different quadrants of the circle or square may be
colored differently to show this

Stomach cancer

Ovarian cancer

Colon cancer

Breast cancer

Figure 1. continued.

C: It's pretty common these days not to know as much about your family, but let's do as much as we can.

Sally: I rang Mum last night, so I know a bit more.

C: Okay, let's start with you and David, what is your date of birth?

Sally: Mine? 29 October, 1964.

C: and your maiden name was?

Sally: Jones.

C: Have you had any serious illnesses yourself Sally?

Sally: No, just well I had bronchitis a lot when I was small.

C: But nothing else?

Sally: Well, how serious do you mean?

C: Have you ever needed to see a specialist or go to hospital?

Sally: Oh no.

C: Now, what about David, what is his date of birth?

Sally: 2 October, 1962.

C: And has David had any serious health problems?

Sally: Nothing I know about, his mother's never mentioned anything, and he was adopted at birth, so no, nothing.

C And Maria, was born on the 4th of May this year?

Sally: Yes.

C: Do you have any other children,

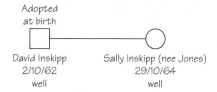

David Inskipp
2/10/62
well

Sally Inskipp (nee Jones)
29/10/64
well

Adopted at birth

Sally: No, she's the first.

C: Did you lose any babies before having her?

Sally: Yes, we'd been trying for years really, I lost three, we were starting to think we'd never have one, I kept losing them, early, couldn't seem to carry past 12 weeks, then it all went okay for her.

C: So you lost three, how far into the pregnancy were you with each one?

Sally: I lost the first at about 10 weeks, that was a shock, we just never expected anything to go wrong... then the second one I was 12 weeks, the last one before Maria I only got to 9 weeks, and they said on the scan it had died a few weeks before.

C: Sounds as though you've had a really rough time

Sally: Yes, and now... we're so worried about Maria.

(counselor and Sally talk for a while about her anxiety)

C: Now can I ask you about your side of the family, do you have any brothers or sisters?

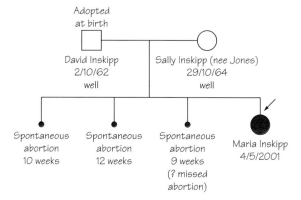

Adopted at birth

David Inskipp
2/10/62
well

Sally Inskipp (nee Jones)
29/10/64
well

Spontaneous
abortion
10 weeks

Spontaneous
abortion
12 weeks

Spontaneous
abortion
9 weeks
(? missed
abortion)

Maria Inskipp
4/5/2001

Sally: Yes, my brother William is 7 years younger than me, he and his girlfriend are getting married soon. They were talking about starting a family.

C: So they don't have any children yet?

Sally: No, Mel always said she'd have to be married first!

C: And did your mother lose any children?

Sally: Well I didn't know this, but last night she told me she lost one before me, it was stillborn, she never saw it.

C: Do you know how far she'd gone, in the pregnancy?

Sally: Yes she said it was full time, a complete shock, no scans then of course.

C: Now, what is your mother's name?

Sally: Joy Jones.

C: And her date of birth?

Sally: 16 July, 1940.

C: Is your mother in good health?

Sally: Yes, she has a bit of blood pressure, but nothing serious.

C: And your Dad?

Sally: He had a kidney stone last year but he passed it, didn't need an operation thank goodness.

C: And his name and date of birth?

Sally: Neal Jones, 12 July, 1940.

In this way, the family tree is built up, gradually moving further back in the family, at least as far back as the client's grandparents if possible. In the course

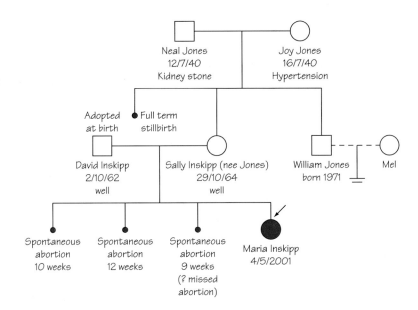

of taking the family history, the genetic counselor also acquires a lot of information about the nature of relationships in the family (such as who talks to whom!) that puts the client into a social context.

4. Dealing with unexpected information when taking the family tree

When taking a family tree because of the family's concern about one genetic condition, it is not uncommon to be given information of significance about another unrelated condition. It is often difficult to know how to deal with this new information, as raising anxiety with no real benefit might cause harm. However, the following questions might help to clarify any action you should take:

- Is the family concerned about the condition?
- Is the family asking questions about the condition?
- Is there likely to be any significant risk to family members?
- Are the risks avoidable?
- Is there any treatment or screening available from which the family could benefit?

If the answer is yes to any of these questions, then undoubtedly the issue should be discussed with the family. Discussion with colleagues is necessary in these situations, to clarify the best course of action.

5. Inheritance patterns

Frequently, the pattern of inheritance of a genetic condition in the family can be identified by looking at the family tree, as each of the patterns have significant characteristics.

5.1 Dominant conditions

If a condition is autosomal dominant, a person who inherits only one faulty copy of the particular gene involved will usually develop the condition [1]. One normal copy of the gene is not sufficient to insure normal cell function. Each child of an affected person has a 50% chance of inheriting the faulty gene from their affected parent. Some examples of dominant conditions are: Huntington's disease, familial adenomatous polyposis, neurofibromatosis, tuberous sclerosis, adult polycystic kidney disease, Marfan syndrome.

Features of a dominant inheritance pattern (*Figure 2*):
- usually more than one successive generation affected;
- people of both sexes are affected;
- male-to-male transmission evident.

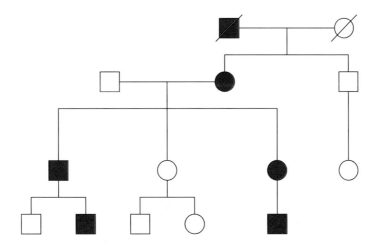

Figure 2. Features of a dominant inheritance pattern.

5.2 Recessive conditions

A person will be affected by a recessive condition if they inherit two faulty copies of the same gene, one from each parent [1]. If both parents are carriers, they each have one faulty and one normal copy of the relevant gene. Carriers are usually healthy, as having one normal copy is enough to insure adequate cell function. Each child of two carrier parents has a 25% chance of being affected with the condition, and a 50% chance of being a carrier. Recessive conditions include cystic fibrosis, thalassemia, sickle cell disease, phenylketonuria, galactossemia, and hemochromatosis.

Features of a recessive inheritance pattern (*Figure 3*):
- usually only one generation affected;
- people of both sexes affected;
- offspring of two normal parents affected.

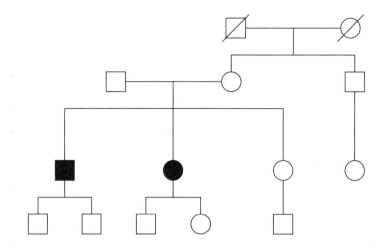

Figure 3. Features of a recessive inheritance pattern.

5.3 *Recessive X-Linked conditions*

Conditions that are known as X-linked are caused by faulty genes on the X chromosome [1]. Females have two X chromosomes, whereas males only have one. Most X-linked conditions are X-linked recessive, which means that a woman will usually not develop the condition, as she will have both a normal and a faulty copy of the gene, and the normal copy usually insures normal function. However, if a male child inherits her faulty copy of the X chromosome he will develop the condition, as he has no normal X chromosome (having inherited a Y chromosome from the father). Each male child of a woman who is a carrier of an X-linked recessive condition will have a 50% chance of inheriting it. Men with an X-linked condition cannot pass it on to their sons, but all their daughters will be carriers. X-Linked recessive conditions include hemophilia, Duchenne muscular dystrophy and fragile X syndrome.

Features of X-linked inheritance pattern (*Figure 4*):
- more than one generation affected;
- males affected more severely than females;
- no male-to-male transmission.

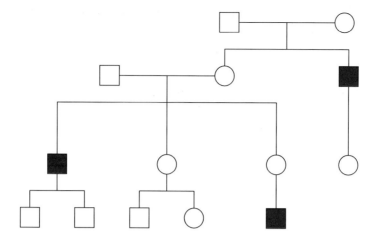

Figure 4. Features of X-linked inheritance pattern.

5.4 X-linked dominant inheritance pattern and Y-linkage

The careful reader will have noted that there are two other possible patterns of inheritance of Mendelian conditions. In practice these are rarely encountered.

In X-linked dominant conditions females will manifest the condition because, despite having two X chromosomes, the presence of just one copy of the mutation is sufficient to cause the condition. However, the expression of the condition may vary because of **mosaicism** due to **X-inactivation** (see section 2.5, p. 54). As only one X-chromosome in any cell is active (switched on) at any time, if the active X has the mutation the faulty gene will be expressed, however if the X chromosome with the normal gene is active the faulty gene will not be expressed. The assumption for males with X-linked dominant genes is that, as all their X-chromosomes have the faulty gene they will express a severe phenotype and in some X-linked dominant conditions this will be incompatible with normal fetal development. You would therefore tend to see an excess of females being born into the family.

If a gene is Y-linked, then the condition would only be expressed in males and would inevitably be inherited by a son from his affected father. It used to be thought that there were no significant disease-causing gene mutations associated with Y-linkage. However, new reproductive technologies such as intracytoplasmic sperm injection are facilitating reproduction for men with low sperm counts. There is a concern that if the sub-fertility was caused by a mutation in one of the genes important for male fertility on the Y-chromosome, then this sub-fertility would be inherited by any male children. As yet this remains a potential concern since babies born using this technique are not yet old enough to reproduce.

6. Making a numerical assessment of risk

When we refer to the possibility of a person developing a particular condition or passing it onto their offspring, we are really talking about *chance*. However, as this chance is known generally in genetics as the recurrence risk, that is the term we use here.

There is a lot of evidence that people do not solely base their genetic decisions upon a numerical recurrence risk value [2]. The individual's perception of the burden of the disease has a huge impact on their decision, and this is a very personal construct based on their own experience of the condition, what they have read or been told, and experience of similar situations [3]. However, a numerical value may help the client, especially if their own assessment of the risk has been unrealistically high or low.

Using the data obtained from the family history and a knowledge of the inheritance pattern, the chance or recurrence risk can be calculated. This is fairly simple in some cases.

6.1 *Case 1: Autosomal dominant inheritance pattern*

William has dominantly inherited adult polycystic kidney disease (APKD). This causes multiple cysts to develop in the kidneys, usually before the age of 30 years [4].

William and his wife are concerned about their two children, Kerry and Harry, and ask about the chance that Kerry and Harry will develop the condition. As William has one normal and one faulty copy of the gene for APKD, and only passes one copy of the gene into each sperm, each child was born with a 50% chance of inheriting the faulty gene (*Figure 5*).

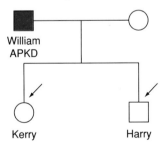

Figure 5. William's family.

6.2 *Case 2: Autosomal recessive inheritance pattern*

Colin has a brother with cystic fibrosis, he asks about his risk of being a carrier of the gene mutation (*Figure 6*).

We know Colin's brother inherited a faulty copy of the cystic fibrosis transmembrane receptor (CFTR) gene from both parents. Both Colin's

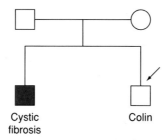

Figure 6. Colin's family.

parents are therefore carriers. Each time they had a child there were four possible combinations of **alleles** (*Figure 7*).

However, Colin does not have cystic fibrosis, so we can eliminate the last possibility from the calculation.

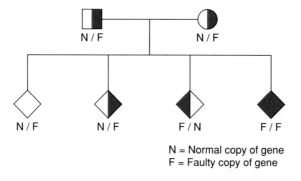

N = Normal copy of gene
F = Faulty copy of gene

Figure 7. Colin's family.

There are three combinations remaining. In two of those combinations a faulty gene is passed on, therefore Colin's chance of being a carrier is two out of three, or a 2/3 chance.

6.3 *Case 3: X-Linked recessive inheritance pattern*

Ruth has two uncles who both died with Duchenne muscular dystrophy (DMD). Ruth's mother has a 50% risk of being a carrier, she had no sons (*Figure 8*).

Ruth's grandmother had two copies of the gene for DMD, one is faulty, the other normal. Ruth's mother was born at 50% risk of being a carrier. Ruth inherited only one copy of the gene from her mother, therefore her chance of being a carrier is 1 in 4 or 25%.

6.4 *Risks estimates based on empirical data*

There are many situations in which the inheritance pattern is not as clear as it is in the examples given above. In some cases, the actual genetic basis of the

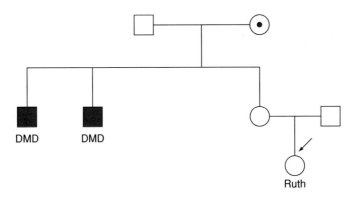

Figure 8. Ruth's family.

condition is still not known, or the condition may be multifactorial. Empirical data are then used to offer the family an idea of the level of risk.

6.5 Case 4: Multifactorial condition

Josie and Fred have had a baby with bilateral renal agenesis (failure of development of the kidney). The baby has not survived. They are very concerned that they may have another child with this same condition.

Renal agenesis is not usually inherited as a **Mendelian** condition, but data show that ~ 5% of couples who have had one child with renal agenesis have another child with the same condition in a future pregnancy [1].

Josie and Fred are told that the risk of a fetus having renal agenesis in a future pregnancy is 5%.

7. Bayesian calculations

The Bayesian calculation is used in genetics for refining the recurrence risk for an individual [5]. The theorem is attributed to Thomas Bayes, an 18th century nonconformist minister who was also said to be an excellent mathematician! The calculation enables us to take into account additional information that might alter the actual likelihood of the person being affected or being a carrier of a condition. Although the calculations may look complicated at first glance, if you just work through them step by step they are actually quite simple. However, unless you are involved in offering genetic counseling, you are unlikely to need to use these types of calculations, so skip *sections 7 and 8* if you feel they are irrelevant to your practice.

The calculation is based on both the prior information (risk based on the family history) and conditional probability relating to information obtained by testing or based on other factors, such as the person's current age. A likelihood ratio is obtained which can easily be converted to a risk estimate.

Step 1
What is the prior probability?
  (a) Risk of being affected/a carrier
  (b) Risk of not being affected/a carrier

Step 2
What is the conditional probability (based on additional information)?
  (A) Chance of this result if affected/a carrier
  (B) Chance of this result if not affected/a carrier

Step 3
Determine the joint probability.
Multiply a × A
Multiply b × B
Express a × A as a ratio of b × B

Step 4
Convert ratio to final probability expressed as a fraction or percentage.

It is easier to demonstrate using an example!

Take the example in case 1 above. William has asked about the risk of his son Harry having APKD. His prior risk is 1/2. However, he has a kidney scan which shows no cysts. Only 2% of those inheriting the faulty gene for APKD will have no cysts at the age of 30 years, so this information can be used to modify Harry's risk.

Step 1
  (a) Prior risk of Harry having APKD = 1/2
  (b) Prior risk of Harry not having APKD = 1/2

Step 2
  (A) Risk of negative scan result if Harry has APKD = 2% = 1/50
       (Because 98% of those affected would have cysts, this is the chance of a false-negative result)
  (B) Risk of negative scan result if Harry has inherited APKD = 1
       (Because all nonaffected will have a negative result)

Step 3
Calculate likelihood of Harry having APKD and of him not having APKD.
Likelihood of having APKD 1/2 × 1/50 = 1/100 = 1/100
Likelihood of not having APKD 1/2 × 1 = 1/2 = 50/100

Likelihood ratio of 1 : 50 that Harry has APKD.

Remember this is a ratio and not a fraction, add both numbers together to obtain the denominator for the fraction.

Harry's chance of having APKD is 1/51 (< 2%).

In case 2 Colin is born with a 2/3 chance of being a carrier of cystic fibrosis. He has a test that detects 90% of mutations found in carriers of cystic fibrosis in our population.

	Carrier	Not a carrier
Prior probability	2/3	1/3
Conditional probability	1/10	1
Joint probability	2/30 = 1/15	1/3 = 5/15
Final probability	1/6	5/6

The chance that Colin is a carrier is 1/6 or about 17%.

In case 3, Ruth has a 25% chance of being a carrier of DMD. However, she has a serum creatine kinase test that indicates her chance of being a carrier is raised. Only 15% of noncarriers would have a CK result in the same range.

	Carrier	Not a carrier
Prior probability	1/4	3/4
Conditional probability	85/100	15/100
Joint probability	85/400	45/400
Final probability	85/130	45/130

The additional information has tilted the balance in favor of Ruth being a carrier of DMD. Ruth has a 65% chance of being a carrier. As it is usually only boys who are significantly affected with DMD, and as she has two possible X chromosomes to hand on, there is a 1 in 4 risk of the fetus having the condition, in each pregnancy. The risk that Ruth will have an affected boy in her next pregnancy is $65/100 \times 1/4 = 65/400$. This translates into a percentage risk of 16%.

There are several problems for you to try at the end of the chapter.

8. Using the Hardy–Weinberg equation to calculate carrier risk

We have seen that it is possible to calculate carrier risk of a particular condition if there is a family history. However, sometimes it is necessary to calculate the population risk of carrying a particular genetic condition. This is the case when a person with no family history partners a carrier of a recessive condition. To calculate the risk to the offspring of this couple, the carrier risk of both partners is required.

The Hardy–Weinberg equation of population genetics can be used for this purpose [6].

The equation states that:

$p^2 + 2pq + q^2 = 1$, where p is the normal allele and q is the mutated allele of a particular gene. If q is the mutated recessive gene, then the frequency of carriers of the recessive gene is $2pq$. In reality, because the population sample is so large, the frequency of the normal allele (p) is regarded as 1. Therefore, for practical purposes the carrier rate in the population (frequency of **heterozygotes**) is $2q$.

Usually the frequency of **homozygotes** (number of affected people) in a population is known, and this figure can be used to calculate carrier frequency.

Take for example a population where 1 in 1600 children are born with cystic fibrosis.

The number of homozygotes with the recessive condition is 1 in 1600, therefore

$q^2 = 1/1600$
$q = $ the square root of $1/1600$
$q = 1/40$
$2q = 1/20$

The carrier risk for someone with no family history of cystic fibrosis in this population is 1 in 20.

This information can be used to assess the risk of a couple having a baby with a recessive condition. The risk is calculated by multiplying

the carrier risk of the father
by
the carrier risk of the mother
by
the chance of the child inheriting the faulty copy of the gene from both parents (1 in 4).

There are helpful web-based resources on both Bayesian and Hardy–Weinberg calculations.

CASE EXAMPLE   CHESTER FAMILY

We can now calculate the risk for a couple where one partner has a family history and the other does not.

Let's return to Colin, who has a brother with cystic fibrosis. Colin's risk of being a carrier of cystic fibrosis is 2 in 3. His partner Julie has no family history of cystic fibrosis. Her risk of being a carrier in this population is 1 in 20.

The chance that they will have a child with cystic fibrosis is:

Colin's risk, i.e. 2 in 3

Multiplied by Julie's risk, i.e. 1 in 20

Multiplied by the chance of a child inheriting both faulty copies of the gene, i.e. 1 in 4

2 in 3 × 1 in 20 × 1 in 4 = 1 in 120.

Of course, if we take Colin's negative cystic fibrosis test result into account his risk is 1 in 6, so the actual risk for this couple is 1 in 6 × 1 in 20 × 1 in 4 = 1 in 480.

9. Lay knowledge and the perception of risk

Lay knowledge or lay belief is a term used to describe the information that exists in the family about the particular genetic condition that affects them [7]. The beliefs are usually rooted in the family's experience of the condition, but may also be based upon general health beliefs, superstition or medical information previously passed on to them. Attributes that are unconnected to the genetic condition (such as hair color) may be linked to the experience of the disease, leading people in the family to believe, for example, that all the redheads in the family will get the condition. In other cases the disease is linked to the sex of those affected, and if the only affected members of a family have all been male, the family may believe that only sons in the family are at risk.

As many people have only a sketchy idea of genetics, they may seek nongenetic explanations for what has happened in the family. Other information about influences on health may be used to draw conclusions about the cause of a disease or syndrome. For example, a mother who has a child with learning delay may question the quality of her diet during the pregnancy [3].

Application of lay knowledge will often lead to particular family members being identified as either being at high risk of the condition or likely to avoid it. This preselection of affected members can liberate those who do not fit the criteria from worry, but may place extra burdens on those predicted to be affected.

For genetic information to be useful for families, they must be able to fit any new information into their family story. It is helpful to use the family's own family tree when explaining the inheritance pattern, so that any discrepancies between the explanation and the lay knowledge held by that family can be addressed. If this is not done, the lay beliefs will persist, as the family's own experience is more powerful than an abstract scientific explanation [3]. Family dynamics can be disrupted if the lay knowledge is challenged by new developments. The family may require time and support to adjust to a new way of thinking about the inheritance of the condition.

In taking decisions, the family will put weight on the burden of the disease as well as the chance of a family member being affected. If the burden of the disease is heavy, even a very low risk may seem oppressive. However, if the family views a disease as mild, they may consider any risk worth taking.

KEY PRACTICE POINT

When explaining an inheritance pattern to the family, use the family's own tree so that the family can integrate the information with their own family experience.

---

10. Conclusion

Drawing an accurate family tree is a skill that can be acquired through practice. The information in the family structure is the basis for risk estimation but other relevant data can be used to refine the risk. However, risk estimation is only one component of genetic counseling and in the next chapter we discuss counseling applications that may be used to support the client.

TEST YOURSELF

Q1. Ben is born with a 25% chance of having inherited polyposis coli. When he has a colonoscopy at 20 years there are no polyps seen. Of those with polyposis coli, 90% would have polyps by 20 years. What is his chance of having inherited polyposis coli?

Q2. Harriet's mother, aunt and sister all had breast cancer under the age of 40 years, therefore Harriet is considered to be at 50% risk of having a breast cancer gene mutation. Her sample is tested for *BRCA1* or *BRCA2* mutations, but no mutation is found. The tests that have been carried out are thought to detect a total of 60% of mutation carriers. What is Harriet's chance of carrying a breast cancer gene mutation?

Q3. In a particular population the number of children born with spinal muscular atrophy (SMA) is 1 per 14 400 children. What is the chance that a person with no known family history of SMA will carry the condition? What is the risk of having a child with SMA for a couple where one partner is known to be a carrier and the other has no family history of SMA?

Q4. Case discussion

Joyce attends a genetic clinic to talk about the family history of cancer. Joyce is a member of a large family with a known genetic mutation. When taking the family tree the genetic counsellor refers to the family tree she has in the notes and comments that the reason the genetics department knows that the gene is in the family is that samples have been looked at from Joyce's sister and her aunt who have had breast cancer. Joyce catches sight of the family tree upside down in the notes and asks why she is shaded in black like her sister and aunt. The counsellor replies that it is because she had breast cancer. At this point Joyce becomes extremely angry and upset, demanding to know who told the genetics department that she had breast cancer; she had never been told she had breast cancer. It emerged that Joyce had had a breast reduction for cosmetic reasons and the family assumed that she had developed breast cancer. Joyce left the clinic still angry with the genetics team and her family and asked for a referral to another genetics centre.

How could this situation have been handled differently?

What does this example illustrate about reported family histories?

References

1. Harper PS (1998) *Practical Genetic Counselling*, 5th edn. Oxford: Butterworth-Heinemann.
2. Lippman-Hand A, Fraser FC (1979) Genetic counselling: provision and reception of information. *Am J Med Genet* **3** (2): 113–127.
3. Skirton H (2001) The client's perspective of genetic counselling – a grounded theory approach. *J Gen Couns* **10(4)**: 311–329.
4. Ravine D, Gibson RN, Walker RG, Sheffield LJ, Kincaid-Smith P, Danks DM (1994). Evaluation of ultrasonographic diagnostic criteria for autosomal dominant polycystic kidney disease 1. *Lancet* **343** (8901): 824–827.
5. Institute of Child Health site for training in molecular genetics. http://www.ich.ucl.ac.uk/cmgs/bayes99.htm
6. Hardy GH (1908) Mendelian proportions in a mixed population. *Science* **28**: 49–50.
7. Richards M, Ponder M (1996) Lay understanding of genetics: a test of a hypothesis. *J Med Genet* **33** (12): 1032–1036.

Further reading

Bennett RL (1999) *The Practical Guide to the Genetic Family History*. New York: Wiley Liss. (An excellent text on taking a family history, with much additional useful information about interpreting the history. Detailed information on familial cancer risk.)

Harper PS (1998) *Practical Genetic Counselling*, 5th edn. Oxford: Butterworth-Heinemann. (Seminal text on genetic counselling, ideal reference work for any health setting.)

Mueller RF, Young ID, Emery AEH (1998) *Emery's Elements of Medical Genetics*, 10th edn. Edinburgh: Churchill Livingstone. (General genetics information related to family history of a range of diseases.)

3

Counseling issues

1. Introduction

Genetic counseling is primarily a communication process [1]. In this chapter, the use of counseling skills to improve communication between professionals and families is described. Several theories of counseling that could be useful when working with families are also discussed.

First, it is important to differentiate between *using counseling skills* and *active counseling*. In any healthcare interaction in which the client has to consider a genetic risk, counseling skills are required [2]. This obviously includes genetic counseling settings, but applies equally to a number of other situations that are far more common in healthcare settings.

Some example situations are:
- A midwife offers a pregnant woman serum screening for **trisomy** 21 risk in the fetus.
- A practice nurse or general practitioner recommends a blood test for cholesterol levels in a person who has a family history of heart disease.
- A breast care nurse discusses routine mammography for screening in a woman with a family history of breast cancer.
- A surgical nurse discusses the benefits and risks of colonoscopy with a man who has a strong family history of colorectal cancer.
- A health visitor does a developmental assessment in a child, because of concerns about the child's delay in reaching milestones.

In each of these situations, the health professional is offering a test that could define the genetic status of the individual concerned (in the pregnant woman, this is of course the fetus). The result of the test may flag up an immediate health concern for that person, but may also give information about long-term concerns, the person's genetic status, and the risks to other members of the family. For example, if the man with a family history of bowel cancer is found to have multiple polyps in the sigmoid colon, he would be considered likely to have inherited a gene mutation for polyposis coli, and the risk to his children of having the same condition is assessed as 50%. It is clear that these issues need to be discussed prior to testing, and the client's feelings about the potential results explored.

For this reason, counseling skills training for almost all health professionals is not only beneficial, but imperative. Many health professionals claim to be 'good communicators,' but anecdotal reports from clients demonstrate that they often feel this is not the case. In one such case, a woman gave birth to a baby with a cleft lip. The midwife kindly said to her 'It's such a pity, she would have been a lovely little girl.' This was probably meant to reassure the mother, but actually left her devastated. In another family, the general practitioner tried to reassure a mother about her son. The 7-year-old boy had recently been diagnosed with neurofibromatosis (NF). The doctor said that she shouldn't worry, because having neurofibromatosis was really no more problematic than having red hair, apparently forgetting that the boy's father had died from an osteosarcoma directly linked with NF.

As clients are often in unfamiliar situations when they meet with health professionals, and may be anxious or concerned, skillful communication is essential if harm to the client is to be avoided. However, competence in health care is based on more than avoiding harm, and holistic health care involves enabling the client to voice their concerns and explore options.

2. The psychological needs of the client and their family

When working with any group of clients in a healthcare setting, the psychological needs of the client must be considered as part of the total package of care for that individual. When the client is facing genetic issues, the situation may be more complex because of the effect of the condition on more than one family member, possibly through many generations. There may be guilt at having passed a condition onto the children, blaming of sections of the family, or secrecy to try to hide the condition. As being affected or at risk of a genetic condition can cause grief in many forms, clients may be in a state of perpetual mourning. For clients who experience this type of psychological pressure, counseling may provide an outlet for emotions that are hidden from the family.

3. Basic counseling skills

In any interaction between health workers and clients, it is obviously desirable that the client feels that their own concerns are heard and addressed, and that they have an opportunity to express feelings connected with their situation. The basic skills required to facilitate this are being able to:
- ask open questions;
- reflect back the client's feelings in a way that affirms them;
- paraphrase and summarize what the patient is saying to 'check you have the story right';
- interpret and use nonverbal communication;
- be comfortable with silences.

Harding Family

Jennifer (nee Harding) is aged 44 years. Her father Cyril was affected with Huntington's disease from the age of 50 years. Jennifer was only 18 when he was diagnosed. It was a huge shock to her; she had not known of the risk to her father or herself. Her father developed paranoia suddenly and was admitted to a psychiatric unit at the local hospital for 4 weeks. Even after he left hospital he was altered, not the father she had known.

Jennifer had always wanted to be a nurse, but after her father became ill she couldn't face working in a hospital. She became a librarian.

When David proposed, she told him that she didn't feel she could risk having children. He knew her father and accepted this decision.

Jennifer dotes on her nieces and nephew. She is now almost too old to have her own family, but at times wonders if she has done the right thing; her yearning for a child of her own is very strong. Cyril died at 62 years after becoming increasingly debilitated. Jennifer finds it so hard to visit her sister Mary, who is now affected, but feels guilty if she doesn't. She is sure Mary would not desert her if the situation was reversed.

Jennifer has lived with grief for many years. After the initial shock of her father's diagnosis, she lived with constant fear of his death. The diagnosis in her sister has awakened all the feelings of fear and distress she felt at the time her father was diagnosed. Jennifer asked whether prenatal testing would be possible if she became pregnant, and was referred to the genetic counselor. On meeting Jennifer to talk about possible prenatal diagnosis, the genetic counselor became aware of her mixed emotions, and made a contract with Jennifer to spend four sessions discussing these issues in a counseling environment.

When the counselor meets Jennifer, she identifies a number of areas of loss connected with the genetic condition in the family. Some of these are:

- *the loss of her father as a supportive parent;*
- *the loss of her chosen career;*
- *the loss of her confidence in her own future health;*
- *the loss of the opportunity to have a family without undue concern;*
- *the death of her father;*
- *the loss of normal sibling relationships after the diagnosis in her sister.*

These tools of counseling will assist the practitioner to hear the client and give them the opportunity to discuss the issues with which they are most concerned. It is not possible to learn counseling via a textbook, as it is a very practical skill that requires both practice in a learning environment and a degree of self-awareness that can be developed within a learning group. However, it is possible to discuss some theories of counseling that may be useful in thinking about clients and the way in which health professionals may be able to support them in adapting to change.

4. Nondirectiveness

The term nondirectiveness has been the focus of much discussion in genetic counseling [3]. It has emerged as a model for practice, partly as a reaction to the criticism that genetics has elements of **eugenics** [4]. The client's choice is considered to be extremely important, and therefore the practitioner avoids directing the client. However, although in principle nondirectiveness seems possible, in practice it has to be acknowledged that very few interactions in health care are completely without some form of directiveness. Pressure can be exerted in many ways, often subtly. It can be implied simply by the wording of certain choices, or the amount of time spent discussing certain options. Think about the phrase

"You can choose to terminate the pregnancy or you can keep your baby."

in this context.

A client may feel coerced into a decision if the counselor says that the majority of people make a particular decision, for example 'It is entirely your decision, but in my experience I find most people want to avoid having a child with Down syndrome.'

However, the philosophy of nondirectiveness does not mean that the counselor leaves the client unsupported in their efforts to make decisions. Enabling shared decision-making is an integral part of the role of a counselor, and requires a degree of skill [5]. For this reason training in counseling skills is necessary.

There are, of course, many situations in health care that require the use of counseling skills. Although many professionals have a great deal of experience in communicating with clients, this is not always sufficient when giving genetic information. The use of the term genetic counselor has been controversial as a professional title because of the resistance of some practitioners to claiming counseling expertise. However, it is clear that in order to fulfill the definition of genetic counseling, skills in this area are required. Counseling skills are used within the session to:

- enable and encourage the client to express their individual concerns;
- provide support to the client through the process of decision-making;
- help the client to adapt to living with the condition or to being at risk of a condition.

4.1 Models of counseling

'Counseling' as an activity differs from the use of counseling skills because in a counseling relationship both parties contract to meet for the purpose of exploring a psychological issue that is of concern to the client [5]. The practitioner should be trained in counseling to the appropriate level, and should be receiving counseling supervision from an appropriately trained and experienced counselor.

There are many models of counseling that are appropriate for use in a genetic counseling setting. Those that are discussed here have been chosen because they can help the practitioner to interpret what is happening for the client, even in a single session. A model provides a framework for understanding the client's difficulties or behavior, however, it is usually not appropriate to vocalize the explanation to the client in the session.

4.2 *Person-centered counseling*

Person-centered counseling was the brainchild of Carl Rogers [7], based on his belief that each person has the ability to solve their own problems and work through difficult situations using their own resources, providing they have support from another person. The person-centered counselor is not 'an expert,' who can solve the client's problems, but rather a supporter whose role is to reinforce the client's self-belief, and enable him or her to explore the situation in a safe emotional environment. Rogers devised the 'core conditions' for a positive counseling relationship, genuineness empathy and warmth.

Genuineness. The counselor is real to him- or herself and to the client. To achieve this, the counselor requires a considerable degree of self-awareness and a belief in the equality of the client.

Empathy. One description of empathy is being able to 'walk in the other person's shoes.' Whereas sympathy involves feeling sorry for the other person, empathy is more connected with trying to understand how the client feels, and communicating that understanding.

Warmth. Understanding the client is not facilitative unless that can be conveyed. The 'gold standard' for the person-centered counselor is the ability to hold every person in unconditional positive regard. Although this itself is a challenge, it helps to reduce value judgments of the client and therefore increases the likelihood that the client will feel free to make the decision that is best for them.

A secondary group of core conditions is helpful in facilitating the client to explore his or her feelings and attitudes, but should only be used when an atmosphere of trust has been established between the client and counselor. These are immediacy, concreteness and challenge.

Immediacy. The use of immediacy in the session helps to reinforce the genuineness of the counselor, and to ground the session in the 'here and now'. The counselor notices and may comment on what is actually happening between the client and the counselor during the session. For example, the client may be clenching her fists, but saying 'I'm not angry.' The counselor may bring this discrepancy to the client's attention by saying, 'You're saying you're not angry, but I notice that your fists are clenched.' This helps the client to become aware of feelings that might be difficult to acknowledge, or of which they may not be consciously aware. Similarly, the client may use immediacy to comment on a client's reactions during the session. For example, the counselor is taking holiday leave, and tells the client that she will not be available for 2 weeks, then notices that the client has become withdrawn. The counselor may comment on this and use it to explore what her being away may mean for that client.

Challenge. Challenge is used when there are discrepancies in the client's account. It is not aggressive, but offered as a way of inviting the client to examine what is happening. For example, the client might say that she had always imagined herself as a mother, then say that it was lucky she didn't have children because she wouldn't have had time for them. This discrepancy in the story could be challenged to help the client become aware of her own feelings.

Concreteness. The counselor uses concreteness to maintain the reality of the discussion and help the client focus on what is currently relevant. For example, a man might be concerned about a diagnosis in his wife, but denies the impact of it on himself. The counselor needs to help him focus on the relevant issues.

Although young babies have a positive self-concept, the process of growing up inevitably damages that concept, and the person loses their confidence in their own ability to direct their lives. Through the use of the core conditions, the counselor aims to restore this positive self-concept, empowering the client. This is a fitting model for use in genetic counseling, where clients frequently have difficult decisions to make, and where the use of person-centered approach reinforces their ability to make those decisions.

4.3 Family systems theory

This model [8] is based on the theory that families work as systems, and changes in one area of the system have an effect on all other components. The system is greater than the sum of its parts, and usually adjusts to changes to function together coherently. However, if a change is not managed by the family, they may need help in 're-framing' what has happened to accommodate the change.

First-order change. When change occurs within a family, it may be first-order change that does not really alter the family arrangement. Family members maintain their relationships with one another in the same way.

Second-order change. Second-order change occurs when the family system really alters as a result of some intervening situation. The family needs to make an adjustment; if this is not possible in a constructive way, the family may need assistance to reframe their relationship, and assign different meanings to behavior, feelings and relationships. Family therapy aims to help the family affect this change.

Family therapy is usually carried out by more than one counselor. One counselor may stay in the room with the family, while the other is observing from another room, via a two-way mirror. At some stage in the session the counselor will leave the room to discuss what is occurring with the observer. Some strategies that may be used to help the family to gain a new perspective are given below.

Reframing. Putting a different tilt on the problem presented by the family, for example, a man who has early dementia is extremely moody. His wife finds it very difficult to cope and complains to the counselor that he is making her sick with worry. The husband's moody behavior may be expressed in terms of the wife's inability to deal with the behavior.

Positive connotation. Noble motivations and responses are ascribed to behavior, and negative labeling of clients is avoided. It is the intention rather than the actual behavior that is relabeled. The wife's responses could be positively ascribed to her love for her husband and her wish for him to be more satisfied with life.

Metaphorical communication. Clients may be encouraged to use fantasy to describe their situation. Counselors may use metaphors to convey concepts, or ask the client to express their situation in sculpture, paintings, or models. One example could be asking the client to use different articles to represent family members, placing them in relation to each other.

Paradoxical directions. The person who is exhibiting the 'problem' behavior may be instructed to perform it at set times and places. For example, a child who overeats may be instructed to eat a particular food on the hour. All of these strategies are used to try and help the family gain a new perspective on the situation and to regain homeostasis.

JENNIFER'S STORY **A FAMILY SYSTEMS PERSPECTIVE**

The family therapist might think about the changes that have occurred for Jennifer and her family. Although Jennifer's sister's diagnosis does not on the surface seem to disrupt Jennifer's life, in reality it is a second-order change as it changes Jennifer's status in the family. She becomes a carer, and in addition the family hopes rest on her as a survivor. The impact of her sister's diagnosis also makes her question her wish for children of her own.

4.4 Psychodynamic theory

Psychodynamic counselors use the client's past history to try to clarify current patterns of behavior [9]. This theory has its roots in Freudian theory, emphasizing the experience of the child as being of importance in establishing recurrent patterns of relating to others.

The relationship the client develops with the counselor is of importance to the therapeutic work. In providing an environment in which the client feels valued and is encouraged to express his or her feelings, the counselor is said to model the 'good enough' parent. Thus the parent/child relationship can be re-experienced in a positive way, enabling the client to take responsibility for his or her own life choices.

As part of the therapeutic process, the counselor may draw parallels between the client/counselor relationship, current relationships for the client and past relationships. This is called the triangle of insight. For example, a client who has difficulty establishing a close relationship with her disabled child may have lost an important person during childhood. The counselor may notice that the client is reluctant to commit herself to a contract with the counselor, fearing dependency. The counselor may interpret this behavior as a recurring pattern of reluctance to become involved, due to fear of sudden loss of the person to whom the client has become attached.

JENNIFER'S STORY **THE PSYCHODYNAMIC PERSPECTIVE**

The psychodynamic counselor may interpret the relationship between Jennifer and herself as important because of Jennifer's loss of her father as a good enough parent during her growing years. The counselor may try to facilitate Jennifer in making her own choices, whilst offering her support, in the way a parent would do with an adolescent child.

The counselor might help Jennifer to relate her current distress about her sister to the fear she felt when her father was diagnosed (past history).

4.5 Transactional analysis

The transactional analysis (TA) model is based on the belief that all humans are born believing in their own worth and the worth of others, but that this belief often becomes damaged during childhood. The belief in your own self-worth is called the *I'm okay* position, whereas belief in the other is termed *you're okay* [10].

Therapeutic counseling is aimed at enhancing the client's belief in themselves, helping them to return to the 'I'm okay, you're okay' state that is necessary for healthy relating. Another aspect of the TA model is the use of the terminology *parent, adult, child*. The three ego states exist in each person, one being dominant at any particular moment in time. The parent ego state is reinforces duty messages such as 'I should.' The child is the more natural, freer state, whereas the adult considers the relative aspects of each course of action and pursues the most logical course.

TA can help to explain why we sometimes react seemingly irrationally with others. For example, a person who we find very dominant can evoke a reaction from our child state. The counselor may try to facilitate the client in responding from the adult state as far as possible. This means becoming aware of one's needs.

JENNIFER'S STORY A TA PERSPECTIVE

> The counselor might help Jennifer to appreciate that the guilt she feels that influences her decision to care for her sister comes from her 'parent' state (I *should* look after her). The desire to have a child may come from her child, with strong input from the parent about denying herself that wish. The counselor may work with Jennifer to help her decide what to do when in the adult mode.

5. Counseling supervision

Any professional who provides counseling support for clients should seek counseling supervision for their work [11]. This is usually provided by an experienced counselor, who is not the person's line manager. The purpose of counseling supervision is to enable the counselor to explore and reflect on their own work with the client. In dealing with emotional areas, issues that impact on the counselor inevitably emerge, and supervision helps to protect both the client and counselor from emotional harm.

Example of counseling supervision in practice

Judy was a midwife counselor working in the antenatal clinic. She was on duty one morning when a woman came into the unit on her way to work because she had not felt fetal movements the previous day. She just 'wanted to make sure everything was all right.' Her husband was away for several days at a conference.

The fetal heartbeat was not seen on scan, and the woman was told her baby had died *in utero*.

Judy arranged to see the woman at home the following week for grief counseling.

When Judy was with the woman, she felt sad for her, but very angry with her husband. She felt unable to offer him any support.

In supervision, Judy puzzled over this, as she generally related well to both mothers and fathers. It was her supervisor who encouraged her to explore any personal issues that might have been influencing her attitude. She realized that she had been very angry with her father for not being present when her mother had died in hospital, and this feeling had been aroused by the circumstances of her patient. When she understood where the feelings had come from, she was able to relate to the bereaved father differently.

6. Conclusion

In all fields of health care, the use of counseling skills helps to enable the client to express themselves, and to make their own choices rather than respond to advice from others. Of course there are situations where health care advice is given appropriately, but in the area of genetics advice-giving is generally not appropriate. Training in counseling skills and ongoing counseling supervision are essential in enabling the professional to care for clients competently.

References

1. Ad Hoc Committee on Genetic Counseling American Society for Human Genetics (1975) Genetic counselling. *Am J Hum Genet* **27**: 240–242.
2. Skirton H (1994) More than an information service? Should genetic services offer clients counselling? *Prof Nurse* **9** (6): 400–402, 404.
3. Kessler S (1997) Psychological aspects of genetic counseling. XI. Nondirectiveness revisited. *Am J Med Genet* **72** (2): 164–171.
4. Kevles DJ (1995) *In the Name of Eugenics*, 2nd edn. New York: Alfred A Knopf/Harvard University Press.
5. Emery J (2001) Is informed choice in genetic testing a different breed of informed decision making? A discussion paper. *Health Expect* **4** (2): 81–86.
6. Egan G (1998) *The Skilled Helper*, 6th edn. Pacific Grove: Brooks/Cole Publishing Company.
7. Rogers CR (1961) *On Becoming a Person*. London: Constable and Co Ltd.
8. Barker P (1998) *Family Therapy*, 4th edn. Oxford: Blackwell.
9. Jacobs M (1985) *The Presenting Past*. Milton Keynes: Open University Press.
10. Hough M (2000). *A Practical Approach to Counselling*. Harlow: Longman.
11. British Association of Counselling and Psychotherapy. Ethical framework – providing a good standard of practice and care. British Association of Counselling and Psychotherapy. http://www.bac.co.uk/members_visitors/members_visitors/htm

Further reading

American Counseling Association website (accessed 2002): http://www.counseling.org/

Berne E (1968) *Games People Play*. London: Penguin Books. (Simple explanation of TA concepts.)

Berne E (1991) *Transactional Analysis in Psychotherapy*. London: Souvenir Press. (Detailed explanation of TA theory.)

British Association for Counselling and Psychotherapy website (accessed 2002): http://www.bac.co.uk/

Cozens J (1991) *OK2 Talk Feelings*. London: BBC Books. (General introductory text on facilitating yourself and others to discuss feelings in a non-judgemental context.)

Egan G (1998) *The Skilled Helper*, 6th edn. Pacific Grove: Brooks/Cole Publishing Company. (The how-to book of counselling skills.)

Elwyn G, Gray J, Clarke A (2000) Shared decision-making and non-directiveness in genetic counselling. *J Med Genet* **37** (20): 135–138. (Discussion on the application of the non-directive approach.)

Harris A (1970) *I'm Okay – You're Okay*. London: Pan Books. (TA concepts explained in very readable form.)

Hough M (2000) *A Practical Approach to Counselling*. Harlow: Longman. (A general text that presents a number of different counselling models and theories.)

Jacobs M (1985) *The Presenting Past*. Milton Keynes: Open University Press. (In-depth discussion of psychodynamic counselling.)

Kessler S (1992) Process issues in genetic counselling. *Birth Defects: Original Article Series* **28**: 1–10. (Discussion on counselling process in context of genetic counselling.)

Kessler S (1997) Psychological aspects of genetic counseling. XI. Nondirectiveness revisited. *Am J Med Genet* **72**: 164–171. (Discussion of application of non-directive principles.)

Mearns D, Thorne B (1988) *Person-Centered Counselling in Action*. London: Sage Publications. (Explanation of use of person-centered counselling.)

Rogers CR (1961) *On Becoming a Person*. London: Constable and Co Ltd. (Rogers' original text on person-centered counselling.)

4 Basic concepts in genetic science

1. Introduction

In the current scientific climate, it is difficult to believe that the double helical structure of DNA was only described as recently as 1953 [1,2], and that the correct number of human chromosomes was finally identified in 1956 [3]. During the course of only one generation, **karyotyping** became a commonplace method of diagnosis of chromosomal abnormalities, and DNA studies were being used to offer families reproductive choices previously denied to them. Since the 1980s techniques in **cytogenetics** and **molecular genetics** have developed dramatically, adding to the information that is now accessible to families. As this new technology assumes more importance in everyday health care, the health professional has to have an understanding of its applications and limitations. The use of genetic technology is already moving from the confined area of medical genetics to almost every other speciality.

It is increasingly apparent that genetic factors influence many aspects of health. The influence of genes on health and disease could be thought of as a continuum. At one end of the spectrum are those conditions that are clearly caused by a single gene **mutation**. This category includes conditions such as Huntington's disease, Duchenne muscular dystrophy and cystic fibrosis. At the other end are the diseases in which inborn genetic mutations do not appear to play a part. Environmental influences are thought to be far more relevant in the causation of diseases in this group, which could include lung cancer and infectious diseases.

When functioning normally, a proportion of genes play an important role in protecting the body from disease. For example, the **tumor suppressor gene** BRCA1 normally inhibits the overgrowth of cells in the breast tissue, therefore reducing the formation of tumors [4]. It is only when these genes are faulty that the disease-inhibiting characteristics are lost. Thus it is not correct to say that a woman has a gene for breast cancer, but rather that she has a breast cancer gene mutation. The role of genes in cancer is discussed more fully in *Chapter 9*.

The aim of this chapter is to provide the health professional with basic scientific literacy to understand the practical application of genetics for families, as described in the rest of the book. The structure of this chapter moves from the larger units of genetic material that are visible with a light microscope (e.g. chromosomes), to smaller units (e.g. genes) that can only be analyzed using molecular genetic techniques. Before reading further, it might be helpful to refresh your memory of the key definitions below.

Key definitions

DNA: Deoxyribonucleic acid. The biochemical substance, which forms the human genome. It carries in coded form the information that directs the growth, development and function of physical and biochemical systems. It is usually present within the cell as two strands with a double helix confirmation. The strands consist of a backbone of sugars and are linked by paired bases. The bases in DNA are thymine (T), adenine (A), cytosine (C) and guanine (G). T and A always pair, as do C and G. This pairing allows for replication of the DNA. The order of the bases specifies the sequence of amino acids, which are the building blocks of proteins. Each amino acid is coded for by a sequence of three bases (codon).

Gene: The fundamental physical and functional unit of heredity consisting of a sequence of DNA. Our genes direct the growth, development and function of every part of our physical and biochemical systems and are the unit by which these characteristics are passed from generation to generation. Genes consist of coding sequences (exons) and noncoding sequences (introns).

Chromosome: The physical structures into which the DNA is packaged within the nucleus of cells. The usual number of chromosomes in humans is 46, 23 pairs. Pairs numbered 1 to 22 are identical in males and females. In addition, females have two identical chromosomes called the X chromosome, males have an X chromosome and a Y chromosome (*Figures 1 and 2*).

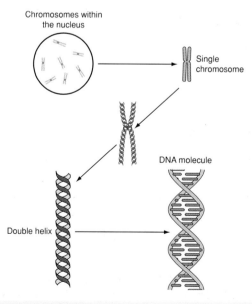

Figure 1. Relationship between a chromosome and a gene.

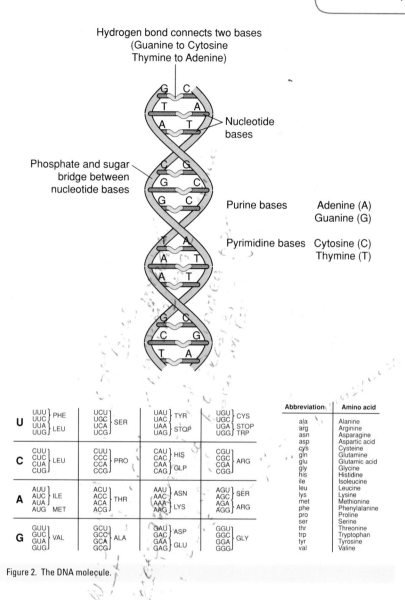

Hydrogen bond connects two bases
(Guanine to Cytosine
Thymine to Adenine)

Nucleotide bases

Phosphate and sugar bridge between nucleotide bases

Purine bases   Adenine (A)
                   Guanine (G)

Pyrimidine bases   Cytosine (C)
                     Thymine (T)

Abbreviation	Amino acid
ala	Alanine
arg	Arginine
asn	Asparagine
asp	Aspartic acid
cys	Cysteine
gln	Glutamine
glu	Glutamic acid
gly	Glycine
his	Histidine
ile	Isoleucine
leu	Leucine
lys	Lysine
met	Methionine
phe	Phenylalanine
pro	Proline
ser	Serine
thr	Threonine
trp	Tryptophan
tyr	Tyrosine
val	Valine

Figure 2. The DNA molecule.

2. The chromosomes

2.1 Mitosis and meiosis

The chromosomes are the packages of genetic material, stored within the nucleus of each cell. Almost every human cell is **diploid**, containing two copies of each of the **autosomes** (chromosomes 1–22), and two sex chromosomes. The exceptions to this are, of course, the germ cells (ova and sperm) in which there is only one copy of each chromosome. These cells are described as **haploid**.

Throughout the human lifetime, cells are produced for new growth or to replace dead cells. Diploid cells are produced by the process of **mitosis** (*Figure 3*).

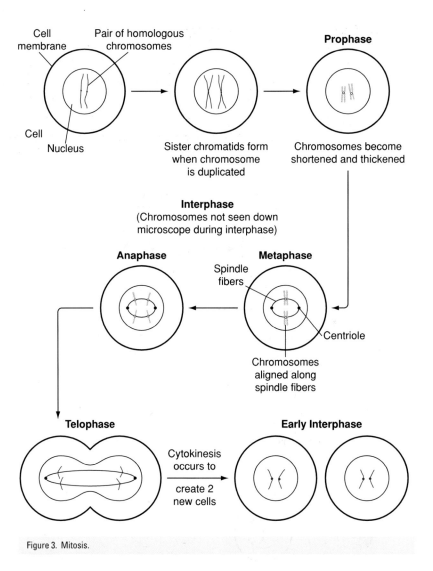

Figure 3. Mitosis.

Within the ovary or testis, haploid cells are made by a different process, called **meiosis** (*Figure 4*).

The first part of the meiotic cycle is similar to mitosis, but during the second meiotic division the diploid cell divides to produce two daughter cells, each containing only one of each pair of chromosomes. If an error occurs during the **disjunction** of the two **chromatids**, then the resultant **gamete** may contain an abnormal number of chromosomes (*Figures 5 and 6*).

2.2 *Chromosomal inheritance*

Meiosis differs from mitosis in two important aspects. The first and obvious difference is the resultant numbers of chromosomes in the daughter cell. This of course is necessary to ensure the embryo contains only two copies of each chromosome. The second is a result of **recombination** (*Figure 7*).

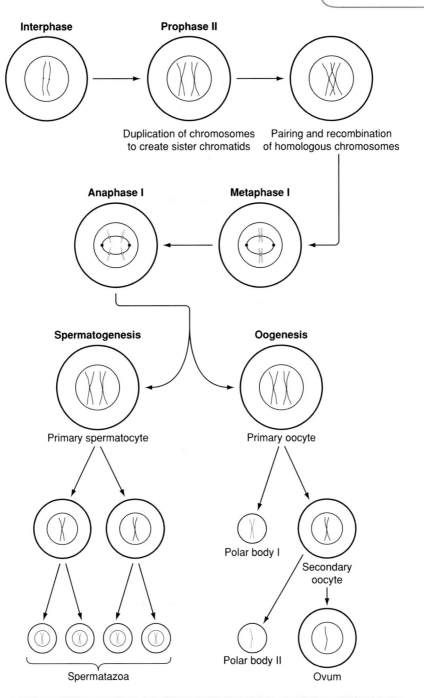

Figure 4. Meiosis.

During the first meiotic division, each pair of chromatids is connected at junctions called **chiasmata** (singular *chiasma*). The chromosome inherited by that individual from their mother is linked with the chromosome inherited from the father. There is some exchange of the chromosomal material

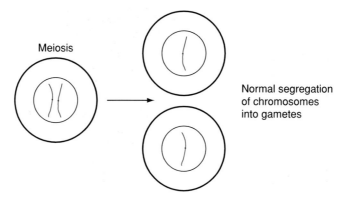

Meiosis

Normal segregation
of chromosomes
into gametes

Figure 5. Disjunction.

between the **homologous** pairs, the resultant two chromatids being a combination of maternal and paternal chromosomes.

The numbers of different combinations are vast, and therefore this phenomenon results in each child being a unique combination of genes derived from all four grandparents.

As already stated, the normal number of chromosomes within a human **somatic** cell is 46. However, the chromosomal arrangement may differ in a number of significant ways that may have implications for the health and development of the person concerned (*Table 1*).

As a general heuristic, changes in the *balance* of the chromosomal material have serious implications for the development of the individual. This usually involves both intellectual and physical developmental changes.

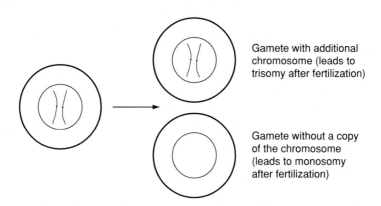

Gamete with additional
chromosome (leads to
trisomy after fertilization)

Gamete without a copy
of the chromosome
(leads to monosomy
after fertilization)

Figure 6. Nondisjunction.

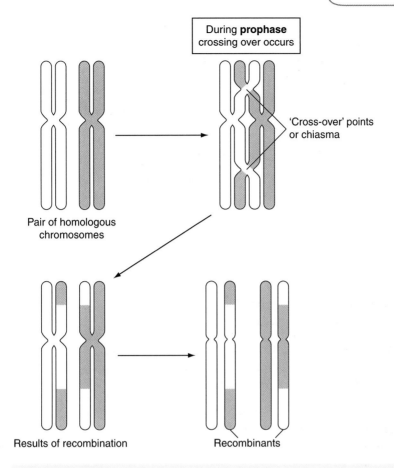

Figure 7. Recombination.

2.3 Imprinting

In some areas of the genetic material, the maternally and paternally derived copies are expressed differently. **Imprinting** is said to occur if the copy from the parent of a particular sex is 'switched off.' For example, the genes that are implicated in Prader–Willi syndrome (PWS) and Angelman's syndrome (AS) are located close to each other on chromosome 15 [5]. The maternal copy of the gene is switched off or imprinted in the PWS region. If a child has no working copy of this gene from the father, they will develop PWS. Similarly, the paternal copy is imprinted in the AS region, and therefore a child requires a copy of the gene from the mother.

2.4 Uniparental disomy

In some circumstances, a child may accidentally inherit two copies of a region of the chromosomal material from one parent. This may apply to a full or

Table 1 Classification of chromosomal rearrangement.

Chromosomal rearrangement	Clinical examples
1. Changes to the total number of chromosomes	
Trisomy	Trisomy 21 Down syndrome Trisomy 18 Edward syndrome Trisomy 13 Patau syndrome
Monosomy	Partial monosomy of autosomes may occur, but complete monosomy of any autosome would not be compatible with life. Cri du chat syndrome is an example of partial monosomy (4p-)
Sex chromosome **aneuploidy**	Turner syndrome 45, X Klinefelter syndrome 47, XXY
Robertsonian translocation (*Figure 8*)	Robertsonian translocation (*Figure 8*) of chromosomes 14 and 15, or 14 and 21 are relatively common
2. Changes in the structure of specific chromosomes	
Balanced reciprocal translocation (*Figure 9*)	There may be an exchange of chromosomal material between any two chromosomes
Deletion	Microdeletion of 22q causes velocardiofacial syndrome
Duplication	Can occur in any chromosome
Inversion: paracentric or pericentric (*Figures 10 and 11*)	Can occur in any chromosome
Insertion (*Figure 12*)	Can occur in any chromosome

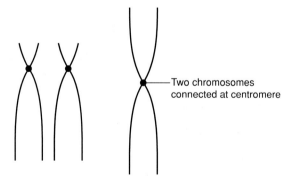

Figure 8. Robertsonian translocation.

partial chromosome. If the uniparental disomy (UPD) includes an imprinted region then a genetic condition may result from the lack of genetic instruction from the relevant parent.

Possible mechanism underlying UPD. During meiosis, a **nondis-junction** event might result in the ovum or sperm having two copies of a

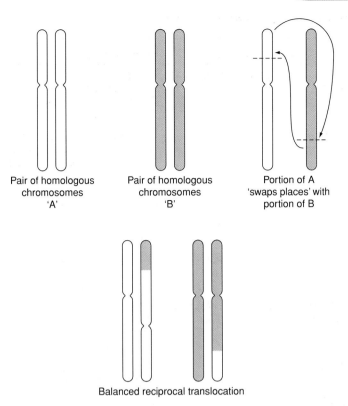

Figure 9. Reciprocal translocation.

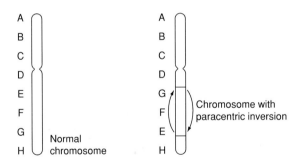

Figure 10. Paracentric inversion

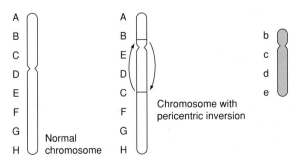

Figure 11. Pericentric inversion

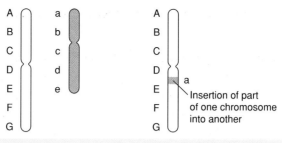

Figure 12. Insertion.

particular chromosome. After fertilization, the trisomy may be corrected by the loss of one copy of that particular chromosome. However, if the copy that is lost comes from the parent that contributed only one copy, the two remaining copies will have originated from the same parent. This will result in uniparental disomy and is called trisomy rescue [6] (*Figure 13*).

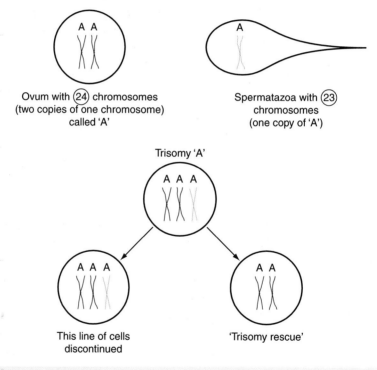

Figure 13. Trisomy rescue.

2.5 X-inactivation

X-inactivation occurs because females have two copies of the X chromosome, while males only have one. If no allowance was made for this imbalance, females would have a double dose of the genes on the X chromosome.

In the embryonic stage, one copy of the X chromosome in a female child is switched off, or made inactive. This usually occurs randomly, with the paternally derived X chromosome being switched off, in approximately 50% of the cells, and the maternally derived X chromosome being switched off in the remainder.

Non random X-inactivation can occur when a female is a carrier of an X-linked condition. In some cases, a greater proportion of the X chromosome that carries the faulty gene will be inactive in comparison to the normal X chromosome. This is sometimes called a skewed X-inactivation. Testing for skewed X-inactivation may be done to detect female carriers. A woman with skewed X-inactivation is likely to be a carrier of an X-linked condition, but unfortunately a normal inactivation pattern does not exclude carrier status so the usefulness of the test may be limited.

CASE EXAMPLE　**HELENA**

Helena is a 24 year old woman whose brother Jim died of Wiskott Aldrich syndrome (WAS), a condition that causes thrombocytopaenia and lowered immunity. There is no DNA available from Jim to test for mutations in the WAS gene. Helena wants to have a baby, and wishes to know if she is a carrier. A study of X-inactivation in her cells shows that in 80% of the cells studied, the X chromosome she inherited from her father is active, while the X chromosome she inherited from her mother is only active in about 20% of cells. The result is strongly suggestive that Helena is a carrier of this X-linked condition.

---

3. Investigating the chromosome structure

3.1 Karyotyping

It is possible for diagnostic purposes to study the chromosome structure of an individual or fetus. In most cases, this is done in a genetics laboratory by a trained cytogeneticist. The chromosome structure is not easily seen during interphase, so cells are cultured in a tissue medium so they can be studied during mitotic division. When sufficient numbers of cells are undergoing metaphase, the process is 'frozen' by adding an agent such as colchicine. This effectively destroys the spindle fibres and the next stage of mitosis (anaphase) is not reached. Without the supporting spindle fibres, the chromosomes spread more evenly around the cell nucleus (*Figure 14*). A salt solution added to the preparation swells the cells, further separating the chromosomes and enabling them to be viewed more easily by the cytogeneticist. The chromosome preparation is 'fixed', placed on a slide, and viewed under the microscope.

A trained cytogeneticist is able to differentiate between the different chromosomes, assessing not only the total number but also the length and structural normality of each chromosome. The use of staining techniques (such as **G-banding**) [7] assists in the differentiation of each part of the chromosome, so that disruptions to the normal arrangement can be more easily detected (*Figure 15*).

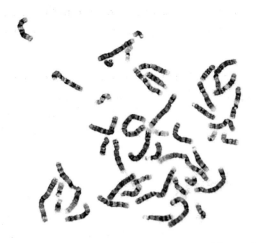

Figure 14. Metaphase spread.

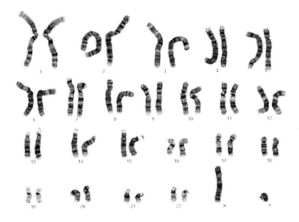

Figure 15. G-banded karyotype.

It is possible for diagnostic purposes to study the constitutional **karyotype**, in situations such as the following:

1. After a couple has had several spontaneous abortions, to detect a possible balanced translocation in one of the parents.

2. When an unbalanced arrangement is suspected in a child who has learning difficulties and/or dysmorphic features.

3. During pregnancy, when fetal cells are obtained via amniocentesis or chorionic villus sampling.

   Chromosome studies can be performed on a variety of tissues. Lymphocytes from a venous blood sample are usually the cells of choice. They are easily obtainable, grow readily in cell culture medium, and provide a good quality chromosome preparation for study. Cells from skin or other tissues (e.g. from a fetus following **stillbirth**) can also be used.

For the purposes of prenatal diagnosis, cells from the chorion or fetal skin cells are used for the culture. Prenatal diagnosis is discussed in detail in *Chapter 6*.

Because of the need to culture the tissues, in many cases the karyotype result will not be available for at least 7 days. The exception is the direct examination of cells from the chorionic villus. Where a rapid result is needed for clinical purposes, for example, for a sick neonate a less detailed result can be made available more quickly.

Chromosome studies may also be requested to aid the classification and management of some cancers, for example, chronic myeloid leukemia. In this situation, the chromosome abnormality will have arisen in the bone marrow as part of the disease process. It is somatic rather than con-stitutional.

3.2 *Chromosomal nomenclature*

When reporting the results of a karyotype, the cytogeneticist always uses a standard nomenclature. The short arm of each chromosome is termed the 'p' arm (for petit) and the long arm is labeled 'q' [8]. A report includes the following details:

- the total number of chromosomes observed;
- the type and number of sex chromosomes;
- any anomaly or abnormality of the chromosomes – a **deletion** is denoted by a 'subtraction' sign, any additional material is preceded by an 'addition' sign.

Using this method, a normal male karyotype will be written as 46, XY and the normal female as 46, XX.

A female child with cri-du-chat syndrome, caused by a deletion on the short arm of chomosome S, will have a karyotype reported as 46, XX, 5p– (*Figure 16*).

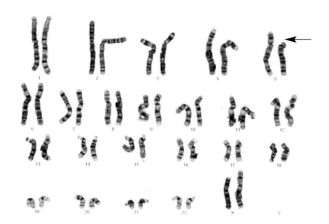

Figure 16. Karyotype of a child with cri-du-chat syndrome. Supplied by Cytogenetics Laboratory, Southmead Hospital, Bristol.

The karyotype of a male child with Down syndrome due to the presence of an additional chromosome will be reported as 47, XY, +21, denoting that the additional chromosome is identified as chromosome 21 (*Figure 17*).

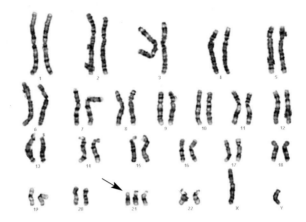

Figure 17. Karyotype of a child with Down Syndrome. Supplied by Cytogenetics Laboratory, Southmead Hospital, Bristol.

3.3 *Fluorescent* in situ *hybridization*

Fluorescent *in situ* hybridization (**FISH**) is a technique (*Figure 18*) used in the laboratory to detect very small deletions or rearrangements in a chromosome. It is a test that is only performed when a specific microdeletion or rearrangement is suspected by the clinician [9,10].

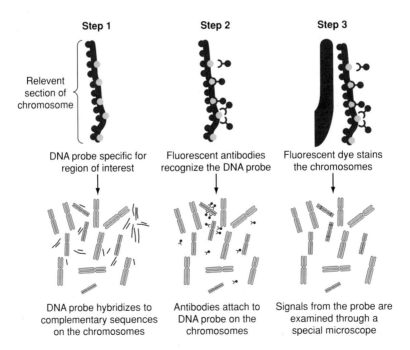

Figure 18. Fluorescent *in situ* hybridization.

A **probe** (i.e. a small sequence of DNA) that matches the normal DNA sequence on the portion of the chromosome to be studied is attached to a fluorescent marker. When the probe is mixed with the chromosome preparation, it adheres to the normal chromosome(s), sending a colored signal. If an autosomal deletion is present, only one signal will be observed, as the probe will fail to attach itself to the abnormal chromosome. Rearrangements can also be detected if the signal is not where it is expected to be.

This technique is frequently used to test for the microdeletion of chromosome 22, known to cause velo cardio facial (or Di George) syndrome [11].

4. Inheritance

4.1 *Mendelian inheritance*

Gregor Mendel, an Augustine monk, solved the riddle of the unit of heredity around the same time as Darwin was developing his theory of evolution. Mendel in his experiments showed that inherited traits are not caused by blending of the characteristics of previous generations but by the inheritance of distinct units of heredity, which we now know as genes. The inheritance patterns in a family tree that are caused by single genes are known as Mendelian.

Mendel's experiments were done with plants. He discovered that if he crossed short and tall pea plants all the offspring were tall. He described tall stature as **dominant** over short stature. He then crossed these tall plants and discovered that about a quarter of their offspring were short. The hidden

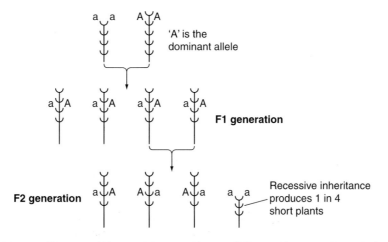

Figure 19. Short and tall peas second generation cross.

recessive gene for shortness was being expressed. He established that genes come in pairs and that if one characteristic is dominant and the other is recessive, the recessive characteristic is not expressed unless both members of the pair of genes are recessive (*Figure 19*). This discovery was made many years before the actual discovery of DNA.

In a species, the single gene responsible for a trait such as stature occupies the same position or **locus** on a chromosome. Each single copy of the gene at the same locus is termed an allele. One copy or allele is inherited from each parent. Within the human population it is possible for there to be many different alleles or versions of the gene at one locus. Within a normal individual there will be a maximum of two different alleles at any one locus. This is because each individual has a maximum of two copies of each gene. This concept will become important to understand when we discuss **linkage** later on in the chapter. These different alleles are simply variations of a normal DNA sequence, they need not be mutations within genes. If an animal or plant carries two identical alleles at a locus they are said to be **homozygous**, if they carry two different alleles at a locus they are **heterozygous**.

One way of thinking about different alleles is to imagine a box of colored balls containing at least two of each color. If you choose two balls you may have two of the same color or two differently colored balls. Selecting two balls of the same color represents homozygosity, whereas the different colored balls represent heterozygosity.

In human genetics there are theoretically five patterns of Mendelian inheritance. In practice only three of these patterns are encountered frequently. These have been described in *Chapter 2*.

4.2 *Expression and penetrance*

The DNA sequence in the gene does not code for a protein directly. The code acts as a template for the production of **messenger RNA** (ribonucleic acid). The mRNA leaves the nucleus and binds to ribosomes where

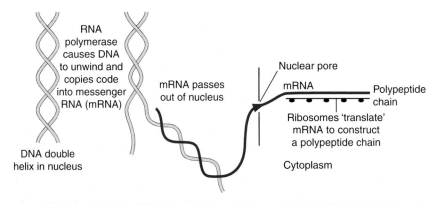

Figure 20. DNA-RNA protein transcription.

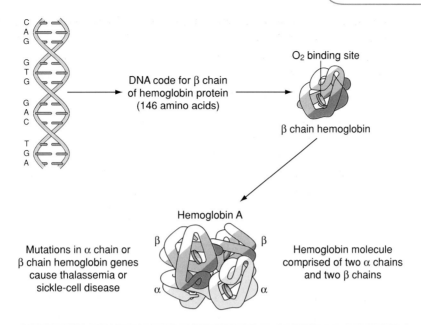

Figure 21. The hemoglobin molecule and globin genes.

it directs the production of proteins. The string of amino acids is built up, as each three base sequence in the mRNA codes for one amino acid (*Figure 20*). The actual physiologically active protein may depend upon the action of several genes, for example, hemoglobin. Although every cell in the body with a nucleus has the same genetic code, not all the genes in every cell are switched on. The genes that are expressed at any one time will depend on the stage of development of the organism and the function of the cell (*Figure 21*).

A gene mutation may not always be obvious within a family tree because it is not fully penetrant. When the family tree is taken it may reveal individuals who must have inherited the gene but who do not manifest the condition.

The inheritance pattern of most of the high-risk cancer gene mutations is autosomal dominant but shows reduced **penetrance**. Although a predisposition for cancer is inherited, some people who inherit a predisposition for cancer will not go on to develop the disease. If a person dies of an unrelated condition before the signs and symptoms of a late-onset genetic disease occur, then the gene will appear to 'skip' a generation. The real explanation may be that the gene is only fully penetrant by old age.

Another explanation for a gene mutation not being obvious within a family tree is variable **expression** at the cellular level. A specific mutation may cause a wide variety of signs and symptoms, even within the same family. This is often most obvious in conditions that affect a number of different body symptoms, for example, Marfan syndrome. Some family members who have the gene have serious complications such as dislocated membranes and a dilated aortic root. Other family members are simply tall and have hyperflexible joints. For this reason a careful clinical examination is necessary before someone is considered not to be at risk [12].

The possibility of variable expression and/or reduced penetrance of a condition should always be taken into account when taking a family history, to insure all family members who are at risk are detected.

5. Genetic mutations

5.1 Types of mutations

Point mutations. Any change in the sequence of DNA within a gene is a mutation in that gene. Many mutations are harmless either because they occur in the noncoding portion of the gene or because they make no difference to the eventual amino acid sequence. Alterations of a single base are called point mutations.

The type of disruption in the sequence of the DNA caused by point mutations. As discussed in previous sections the DNA code is read in sequences of three bases, each three base codon coding for a single amino acid.

CGGGTTTTGAAGCCGGGC
| CGG | GTT | TTG | AAG | CCG | GGC |
| Arg | Val | Leu | Lys | Pro | Gly |

To illustrate the point we use a simple well-known English sentence.

The Cat Sat On* The Mat

(N.B. On* is used to simply give all words three components)

If there is a change in one of the letters the way the sentence reads is affected in different ways depending on the change.

A point mutation would be the substitution of one letter for another, e.g.

The	Cat	Sat	On*	The	*Rat*
CGG	GTT	TTG	AAG	CCG	AGC
Arg	Val	Leu	Lys	Pro	Ser

This change still allows the sentence to be read although the sense is changed slightly. This would be equivalent to a missense point mutation. It may or may not cause a disruption to the final protein structure. Most proteins can tolerate some change to their amino acid sequence.

The	Cat	**Smt**	On*	The	Mat
CGG	GTT	**TAG**	AAG	CCG	GGC
Arg	Val	*STOP*			

This change has disrupted the sense of the sentence and would be equivalent to a nonsense mutation. Nonsense mutations cause the translation of the messenger RNA to end prematurely resulting in a shortened protein. In most cases this happens because the mutation causes the codon to be read as a stop codon. Nonsense mutations usually have a serious effect on the encoded protein and may cause a mutant phenotype.

Insertion or deletion of one or more bases may also cause mutant phenotypes. The effect of these depends on whether the number of bases involved is a multiple of three or not. If it is a multiple of three the reading frame of the DNA sequence remains constant although there will be extra or missing amino acids. The exact phenotypic effects of these in frame mutations will depend on the effect of the mutation on the structure of the encoded protein.

The	Sat	On*	The	Mat		
CGG	TTG	AAG	CCG	GGC		
The	*Fat*	Cat	Sat	On*	The	Mat
CGG	GTT	GTT	TTG	AAG	CCG	GGC

If the number of bases inserted or deleted are not multiples of three then the reading frame is disrupted and the sequence of amino acids downstream from the mutation will be read differently. These frameshift mutations usually have a serious effect on the eventual protein.

The	CaS	atO	n*T	heM	at	
CGG	GTT	TGA	AGC	CCG	GC	
Tth	eCa	tSa	tOn	*Th	eMa	t
CCG	GGT	TTT	GAA	GCC	GGG	C

Mutations that have no effect on the encoded protein and therefore do not result in a mutant phenotype or that occur in the noncoding portion of the genome are called **polymorphisms**. They tend to accumulate in the DNA of organisms and contribute to the variability between individuals.

Gross mutations. Large sections of DNA can also be altered and these mutations normally disrupt the amino acid sequence. Deletions are a loss of a portion of DNA which may be just part of the gene or the entire gene sequence. There may be sequences of DNA inserted either from another portion of the genome or as **duplications**. A novel mutation responsible for a number of human diseases such as Huntington disease and Fragile X syndrome is an **expansion** of triplet repeats [13,14]. Some portions of the genome consist of a series of repeats of bases. An expansion of the normal number of repeats can cause a mutant phenotype.

Normal sequence

CGGGTT**CAGCAGCAG**AAGCCGGGC

Mutant expanded sequence

CGGGTT**CAGCAGCAGCAGCAGCAGCAGCAGCAG**AAGCCGGGC

It is important to remember that although genetic mutation can cause human disease, random mutation has contributed to the development of the myriad of species on the earth today.

5.3 Molecular tests

Molecular genetic laboratories use a variety of different techniques to detect mutations or to track genes through families. The basis of much genetic testing currently in clinical genetics is the **polymerase chain reaction (PCR)**. With PCR, specific DNA sequences can be copied many times to yield large quantities of the particular portion of DNA corresponding to genes or fragments of genes. These can then be analyzed or manipulated. The exact method of analysis used depends upon the characteristics of the DNA sequence of interest and the type of potential mutations. The technique of PCR has also been important in the development of forensic DNA analysis. It has allowed the amplification of small amounts of DNA to provide a unique DNA fingerprint, which can be used as evidence [15].

Polymerase chain reaction. Amplification of the target DNA sequence occurs through repeated cycles of DNA synthesis. In clinical genetics the usual source for the DNA template used for the reaction is DNA extracted from cells from a patient. As in karyotyping, this can be any nucleated cell, but commonly a blood sample is used. DNA can also easily be extracted from cells in the saliva or scraped from the inside of the cheek (buccal smear). This technique can also be used for prenatal diagnosis. Although DNA analysis is possible on cells harvested from amniotic fluid, the preferred sample is a chorionic villus biopsy, as this yields greater amounts of DNA.

In addition to the target DNA, the other components of a PCR are:
- the primers – short sequences of single stranded DNA that bind by complementary **base pairing** either side of the sequence containing the gene of interest;
- DNA polymerase – an enzyme that copies the target sequence and that acts at a specific temperature;
- molecules corresponding to the four bases – these are used in the synthesis of new DNA strands.

In a PCR, the template DNA (usually all the chromosomal DNA) is put into solution with the primers, the DNA polymerase and the bases. The reaction is heated which destabilizes the double helical structure of the template DNA. This separates into two single strands (denaturing).

The reaction is then cooled to a temperature that allows the primers to bind to the single-stranded DNA without allowing the double helix to reform (annealing). It is then heated to a temperature that allows the DNA polymerase to become active. The polymerase copies the sequence of the template DNA starting at the primers and using the bases present in the reaction for synthesis of new single DNA strands (synthesis).

This cycle is repeated 20–40 times dependent on how much of the target sequence was present in the original DNA template. In the first cycle synthesis carries on beyond the end of the target sequence because there is nothing to stop it. Over subsequent cycles the newly synthesized strands, which end with a primer sequence, themselves act as templates and eventually only the target sequence is amplified (*Figure 22*).

Once sufficient quantities of target DNA have been generated, it can be used for clinical analysis or research. Techniques used to identify specific known gene mutations depend on the nature of the mutation. Analysis of specific mutations often utilizes DNA probes that **hybridize** to the mutation. DNA probes are single-stranded specific DNA sequences; they are radioactively or fluorescently labeled. If a probe is specific for a mutation then it will show a signal if the mutation is present and will not show a signal if it is absent. Probes may designed to detect either normal or mutated sequences.

Southern blotting. Southern blotting (named after its inventor Ed Southern) was the first method of analysis of DNA that utilized **hybridization** (16). The technique can be used on large fragments of DNA, which are too large to be amplified in a PCR (*Figure 23*).

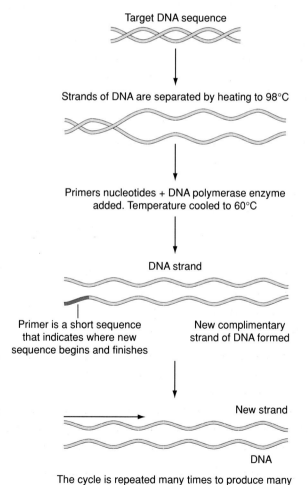

Target DNA sequence

Strands of DNA are separated by heating to 98°C

Primers nucleotides + DNA polymerase enzyme added. Temperature cooled to 60°C

DNA strand

Primer is a short sequence that indicates where new sequence begins and finishes

New complimentary strand of DNA formed

New strand

DNA

The cycle is repeated many times to produce many identical strands of the DNA sequence of interest

Figure 22. PCR.

Genomic DNA is obtained from a sample of tissue (usually blood)

A **restriction enzyme** is used to cut the DNA into small fragments

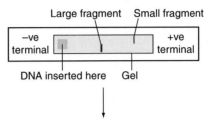

The fragments are subjected to electrophoresis. Smaller fragments
move further along the gel than larger ones

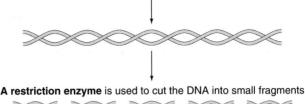

Large fragment   Small fragment

−ve
terminal

+ve
terminal

DNA inserted here   Gel

DNA fragments transferred onto a filter sheet.
Filter sheet is put into a bath of radioactive probes
After the probes have attached to specific DNA fragments, the filter
sheet is exposed to X-ray creating an autoradiograph or *autorad*

Figure 23. Southern blotting.

The DNA is cut into fragments using restriction enzymes that only cut at specific sequences of bases, for example, after a CTG sequence. The DNA is extracted from the cell and cut into fragments. Because the restriction enzyme cuts at specific sites a number of different sized fragments are produced, these are called **restriction fragment length polymorphisms (RFLP)**. Their length may be altered by a mutation. The fragments are separated by size by placing the reaction on an electrophoretic gel. Because the fragments are slightly negatively charged they move from the negative end of the gel to the positive end. The larger fragments do not move as far as the smaller fragments and therefore the large ones are near the top of the gel, the small ones at the bottom. The fragments are then transferred to a membrane by blotting. A labeled DNA probe is hybridized to the membrane. If the probe is radioactively labeled, the membrane is placed next to an X-ray file and the film is developed. After several hours the film will reveal a number of bands corresponding to the fragments to which the probe has hybridized. A specific mutation can be detected if it alters a restriction enzyme cutting site, so that the presence of a mutation causes a different band pattern to the absence of a mutation. Large deletions or insertions also change the length of the restriction fragments.

Southern blotting has also been used extensively in research to isolate genes and track them through families. Southern blotting analyzes DNA, another similar technique (Northern blotting) is used to analyze RNA.

A direct test for a mutation can be performed if:
- A specific mutation is known;
- There is either a specific probe for its presence or absence;
- If the mutation alters a restriction enzyme cutting site or the size of a band.

For some genetic diseases the mutant phenotype can be caused by a variety of different mutations in the same gene. If the exact mutation is not known, genetic testing with a family may be possible using linkage.

Linkage. Linkage is the tendency for alleles whose loci are located close to each other to be passed on together on the same chromosome (*Figure 24*). The closer together on a chromosome the alleles are, the more likely they are to be inherited together. If alleles are further apart on a chromosome, because of recombination during meiosis the chance of them being inherited together is reduced (see recombination).

The phenomenon of linkage has been used to map the location of genes onto chromosomes. Prior to the identification of specific disease-causing mutations, it could also be used in families for genetic testing as below.

The use of linkage requires:
- that the family is sufficiently informative, that is, there are enough affected and unaffected members from whom samples can be obtained;
- that the gene of interest has been mapped to a specific chromosomal location;
- that the allele that is linked to the gene is sufficiently variable (or polymorphic) for the inheritance to be determined.

The allele linked to the gene is called a genetic marker. Linkage techniques using a variety of polymorphic genetic markers have been used to map the human genome (*Figure 25*).

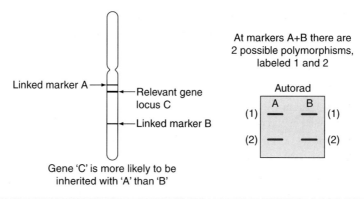

Figure 24. Linkage.

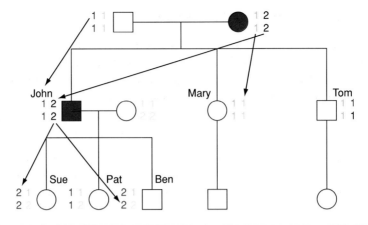

Figure 25. Linkage study (pedigree with haplotypes).

Linkage depends upon the fact that there is considerable variation in the human genome between individuals and between homologous chromosomes within individuals. Molecular genetic techniques are being continuously developed utilizing this variability for the analysis of genes and for examining differences between individuals.

Polymorphic DNA markers

RFLPs are only one type of polymorphism used to detect mutations or to clarify individual differences at the DNA level. DNA probes can also be used to detect **variable number tandem repeat (VNTR) polymorphisms.** Within genomic DNA are a number of sequences consisting of repeated units. The precise number of repeats at a specific locus varies between and within individuals. Analysis of VNTR polymorphisms requires digestion of the DNA sample by a restriction enzyme that recognizes sequences flanking the repeat unit. Because the number of repeats within that section of DNA is variable a variety of fragment sizes will be created. If many probes are used a unique DNA fingerprint of an individual can be built up.

Analysis of VNTR polymorphisms requires radioactivity and the fragment sizes are too large to amplify well with PCR. The standard tools for PCR linkage analysis are smaller repeat units, mostly tri- or tetranucloetide CA repeats.

Recently developed polymorphic markers are **single nucleotide polymorphisms (SNPs)**. Technology is being developed to allow multiple testing for the presence or absence of a SNP utilizing a few thousand primers and a DNA chip. This will allow simultaneous analysis of thousands of markers across the whole genome.

5.4 *Methods of mutation scanning*

In some genes there may be many possible disease-causing mutations and genetic testing requires elucidation of the mutation that is relevant for that particular family. This may be undertaken in the genetic service laboratory but there are few quick cheap and reliable methods for doing this. Much mutation searching is carried out within the context of research laboratories and may only be possible during the lifetime of a particular research project.

There are a number of different techniques, such as sequencing, protein truncation testing, mismatch cleavage and heteroduplex gel mobility. If you are interested in more detailed information, see the text by Strachan and Read (listed in Further Reading).

Many DNA tests are complex and the laboratory will need to work with samples from a particular family for some months before a test is ordered for prenatal diagnosis or predictive testing.

6. Polygenic inheritance

If a characteristic is controlled by a single gene, a change in that gene will produce a distinct phenotype (a specific group of features or abnormalities). If one working allele is sufficient for normal activity, the dominant phenotype is seen in the homozygotes (two faulty genes) and heterozygotes (one normal and one faulty gene). The recessive phenotype is seen in homozygotes where the functional allele is missing (and there is therefore no working gene). For some Mendelian traits such as blood groups there is more than one dominant allele but there are still several distinct phenotypes. The genotype makes a large contribution to the phenotype.

However, for many human characteristics such as height, intelligence or skin colour there is continuous variation usually following a normal distribution curve. Although there may be exceptional single alleles that cause extremes of variation, for example single genes leading to learning disability or extreme short stature, the trait is controlled by more than one gene and environmental action is also important. For example, children tend to be about as tall as their parents, but a starved child would not be as tall as their well-nourished brothers and sisters. These continuous traits are said to be **polygenic** or **multifactorial** and are under the control of many genetic loci in interaction with the environment.

There are other traits which do not vary continuously but where the control of the trait is due to many genes as well as environmental effects. The concept of a threshold effect is used to explain this. A trait such as a neural tube defect (**spina bifida** or **anencephaly**) is either present or absent. There are hypothesized genetic factors and known environmental factors that contribute to the risk. A combination of genotype, environment and other chance factors push some conceptions over the threshold from a fetus without a neural tube defect to one with. Diabetes and cancer are human diseases where there are known alleles that contribute to risk but again there has to be sufficient combination of risk factors to reach the threshold for disease.

7. Mitochondrial inheritance

So far, we have been concerned with the DNA within the nucleus of the cell. Within the cytoplasm of the cell the mitochondria contain their own DNA (*Figure 26*). The **mitochondrial** genome is circular and contains much less DNA than the nuclear genome [17]. Some of the proteins in mitochondria are under the control of the mitochondrial genome, but many are under the control of the nuclear genome. The unusual feature about mitochondria is that they are inherited from the mother but not the father. A woman will pass on her mitochondria to her sons and daughters, her daughters will pass on their mitochondria, her sons will not. This is because the oocyte contains mitochondria within the cytoplasm. The sperm passes on the nuclear DNA to the embryo but not cytoplasmic DNA, because the mitochondria are in the tail of the sperm which does not contribute to the fertilized ovum. Certain inherited human disorders show maternal inheritance and have been shown to be due to mitochondrial mutations. The mitochondrial gene sequence has been established. Mitochondrial DNA sequences and mutations have also been used to examine evolution and population movements as they provide an unbroken maternal link to our ancestors.

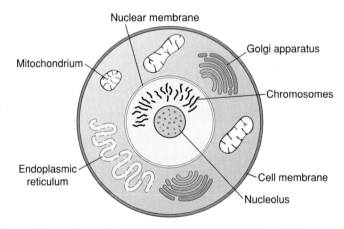

Figure 26. The cell including the mitochondria.

8. The Human Genome Project

The Human Genome Project is a unique international collaboration to determine the entire nucleotide sequence of the human genome. It originally arose because of the need to develop new mutation detection methods. The first draft has now been published [18,19] and has been the subject of much media interest. The newer techniques for mapping genes using genetic polymorphisms as outlined in this chapter have contributed to this project. The final product will be a map. It will provide a guide for further exciting research into the control of gene regulation, expression and function. The single gene disorders, which will theoretically be the easiest target for the development of novel therapies, are rare. Most common disorders are multifactorial.

Therefore, caution should be exercised in predicting how soon important medical advances will be realized. However, with the sequencing of the human genome, the map and the tools are in place to embark on the research.

9. Pharmacogenetics

Pharmacogenetics targets drug therapy to those patients for whom a particular medication will have the greatest treatment effect or least side effects [20]. Genes control the production of enzymes that metabolize drugs. In Chapter 2, it was explained that all humans have genetic variation. If a variation in the usual sequence of base pairs occurs in over 1% of the population, this is called a polymorphism, and is considered to be a normal variant. The normal variance in some genes influences the effect of certain medications. If a patient has certain polymorphic variations in these genes, this will influence the way that certain drugs are metabolized in their body.

Traditionally, medicines were developed to treat signs and symptoms of a condition. Where different medications were available, the patient tried these until a satisfactory option was found. This of course means that a proportion of patients will be treated with drugs that have no beneficial effect for them, or that produce unwanted side effects or complications.

Testing a patient for genetic variation prior to prescribing a drug will enable the right drug to be administered in the right dose. A simple genetic test to look at the polymorphic variation present in the patient will provide the information needed to target the drug. For example, some patients have difficulty metabolizing beta receptor antagonists, because of lack of function of the cytochrome P450 enzymes. Using pharmacogenetics, these patients could be identified and toxicity from the drug avoided.

10. Gene therapy

Although much is hoped in terms of **gene therapy**, at present such treatment is still in the realms of research rather than clinical service [21]. The aim of gene therapy is to introduce a functioning copy of a gene into the body of the patient, so that it is able to function and repair the damage done by a faulty copy of the gene. Even where making a normal copy of the gene is possible, transporting it into the appropriate cells is more difficult. Some research involves inserting the gene into a virus, to enable it to be passed into cells. For example, gene therapy for cystic fibrosis involves introducing a normal copy of the gene into the tissues of the lung via a nebulizer. The therapy needs to be repeated at intervals, and is not a lasting solution or cure. However, as techniques improve, this situation may improve. Effective gene therapy is still many years away for most patients affected with a genetic disease. Other applications of gene therapy may provide treatments for cancer and infection.

11. Conclusion

The following chapters illustrate the application of these genetic technologies to real-life situations in which families find themselves in. It is essential that healthcare professionals are able to understand the implications of these technologies for the families with which they are in contact. Only with this understanding will they be able to interpret both the limitations and the possibilities of these scientific advances.

TEST YOURSELF

Q1. Can you think of a possible genetic explanation for the higher incidence of trisomy in children of women who are over the age of 40, when compared with the offspring of younger mothers?

Q2. Describe the change to the normal human chromosome pattern designated by the following:

(a) 45, X
(b) 47, XY, +18
(c) 47, XY, +21

Q3. How would the following be described in a cytogenetic report, using ISCN nomenclature?

(a) A normal female chromosome arrangement.
(b) A chromosome arrangement indicative of Klinefelter syndrome.
(c) A chromosome arrangement indicative of Patau syndrome.

Q4. What is the function of DNA?

Q5. Samples from the family below are taken and used in a linkage analysis for research. The following results were obtained from the markers A B C D and E.

What possible explanation would account for this and how would you explain it?

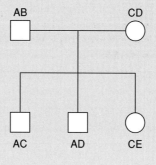

References

1. Watson JD, Crick F (1953) Molecular structure of nucleic acids: A structure for deoxyribose nucleic acid. *Nature* **171** (4356): 737–738.
2. Watson JD, Crick F (1953) Genetical implications of the structure of deoxyribonucleic acid. *Nature* **171** (4361): 964–967.
3. Tijo H, Levan A (1956) The chromosomes of man. *Hereditas* **42**: 1–6.
4. Andres JL, Fan SJ, Turkel GJ *et al.* (1998) Regulation of BRCA1 and BRCA2 expression in human breast cancer cells by DNA-damaging agents. *Oncogene* **16** (17): 2229–2241.
5. Driscoll DJ, Waters MF, Williams CA *et al.* (1992) A DNA methylation imprint, determined by the sex of the parent, distinguishes the Angelman and Prader–Willi syndromes. *Genomics* **13** (4): 917–924.
6. Purvis-Smith SG, Saville T, Manass S *et al.* (1992) Uniparental disomy 15 resulting from correction of an initial trisomy 15. *Am J Hum Genet* **50** (6): 1348–1349.
7. Seabright M (1971) A rapid banding technique for human chromosomes. *Lancet* **2** (7731): 971–972.
8. ISCN (1995) *An International System for Human Cytogenetic Nomenclature.* Basel: Karger.
9. Trask BJ (1991) Fluorescence *in situ* hybridization: applications in cytogenetics and gene mapping. *Trends Genet* **7** (5): 149–154.
10. van Ommen GJ, Breuning MH, Raap AK (1995) FISH in genome research and molecular diagnostics. *Curr Opin Genet Dev* **5** (3): 304–308.
11. Scambler PJ, Kelly D, Lindsay E *et al.* (1992) Velo–cardio–facial syndrome associated with chromosome 22 deletions encompassing the DiGeorge locus. *Lancet* **339** (8802): 1138–1139.
12. Gray JR, Bridges AB, West RR *et al.* (1998) Life expectancy in British Marfan syndrome populations. *Clin Genet* **54** (2): 124–128.
13. The Huntington's Disease Collaborative Research Group (1993) A novel gene containing a trinucleotide repeat that is expanded and unstable on Huntington's disease chromosomes. *Cell* **72** (6): 971–983.
14. Fu YH, Kuhl DP, Pizzuti A *et al.* (1991) Variation of the CGG repeat at the fragile X site results in genetic instability: resolution of the Sherman paradox. *Cell* **67** (6): 1047–1058.
15. Jeffreys AJ, Wilson V, Thein SL (1985) Individual-specific 'fingerprints' of human DNA. *Nature* **316** (6023): 76–79.
16. Southern EM (1992) Detection of specific sequences among DNA fragments separated by gel electrophoresis. *Biotechnology* **24**: 122–139.
17. Anderson S, Bankier AT, Barrell BG *et al.* (1981) Sequence and organization of the human mitochondrial genome. *Nature* **290** (5806): 457–465.
18. Venter JC, Adams MD, Myers EW *et al.* (2001) The sequence of the human genome. *Science* **291** (5507): 1304–1351.
19. Lander ES, Linton LM, Birren B *et al.* (2001) Initial sequencing and analysis of the human genome. *Nature* **409** (6822): 860–921.
20. Ensom MH, Chang TK, Patel P (2001) Pharmacogenetics: the therapeutic drug monitoring of the future? *Clin Pharmacokinet* **40** (11): 783–802.
21. Scollay R (2001) Gene therapy: a brief overview of the past, present, and future. *Ann NY Acad Sci* **953**: 26–30.

Further reading

Gardner A, Howell RT, Davies T (2000) *Human Genetics*. London: Arnold. (A very helpful basic book for those who require clear simple explanations of the genetics and laboratory techniques.)

Passarge E (1995) *Color Atlas of Genetics*. New York: Thieme. (Diagrammatic explanations, quick reference.)

Strachan T, Read AP (1999). *Human Molecular Genetics*, 2nd edn. Oxford: BIOS. (Very detailed account of principles of human genetics.)

5 Pre-conception

1. Introduction

The aim of pre-conceptual care is to identify situations in which the parents (particularly the mother) or the fetus may be at any additional health risk, and to take steps to minimize that risk before pregnancy if possible. In some cases, the risk cannot be altered, but the parents have the opportunity to absorb the information relating to their situation and to think about potential options available to them.

Pre-conceptual care and counseling for families who are planning a family can be beneficial for several reasons:
- it enables high-risk situations to be identified prior to pregnancy occurring, this usually gives the couple more possible options for action;
- it enables the couple to think about difficult decisions before they have an emotional investment in a current pregnancy;
- it enables the healthcare team to be prepared for additional care or possible testing during a pregnancy.

Although dedicated 'pre-conceptual clinics' are not yet widely used by couples planning a pregnancy, there are a great number of opportunities that exist for health professionals to offer information and advice that might alter the outcome of a future pregnancy. Professionals who may routinely encounter situations in which pre-conceptual advice may be offered include:
- nurses or doctors working in family planning clinics, who will meet women who are currently using contraception but may wish to start a pregnancy within the next few years;
- health visitors, who are caring for families after the birth of one child and who may wish to increase their family;
- midwives, following the birth of a baby;
- general practitioners, in the course of routine health care of the family;
- fertility clinic doctors and nurses, as couples seeking advice will be hoping to achieve a pregnancy;
- nurses working in gynecology or genitourinary clinics;
- practice nurses, for example, when undertaking cervical smear tests;
- school nurses, when providing health education.

Obviously, in each of these situations the relevance and opportunity to offer pre-conceptual counseling will vary, but an awareness of the possibility of offering information prior to pregnancy should not be overlooked. General health education that may improve the outcome for the fetus and mother can be offered by any relevant health professional.

This should include:

- general information about a healthy diet;
- promotion of regular exercise;
- advice on restriction of alcohol use in pregnancy and advice on cessation of cigarette smoking prior to conception, because of the proven adverse effects on fetal growth;
- information about the use of medication or dietary supplements (including vitamins) during pregnancy. Many women will, for example, be unaware of the potential for toxicity from overdose of fat-soluble vitamins;
- a review of prescribed medication being used by the mother to assess its safety in pregnancy;
- advice on the use of folic acid prior to conception (discussed further later in this chapter);

Cypros Family

George and Anna are newly married. They both come from families who originated in Cyprus.

George and Anna knew nothing about thalassemia when they married, but when they were on holiday in Cyprus (a gift from Anna's father), they met Anna's distant relations. One of Anna's cousins, a young man called Michael, has beta thalassemia, and Michael's mother told them the disease was inherited.

Back home in the UK, Anna and George ask about thalassemia when they next go to the family planning clinic for Anna's oral contraceptive pill. The nurse offers to refer them to the sickle cell and thalassemia center. The thalassemia counselor explains the condition to them, and offers them carrier testing. They both agree that they would rather know now if their future children are at risk, and have blood taken. Deep down, George does not believe there is any cause for concern, as there is 'nothing like that in his family'. Anna has the test 'just to give her peace of mind'.

Four weeks later George and Anna return to the thalassemia clinic and are told that they are both carriers of beta thalassemia. They are stunned, and hardly take in what they are being told. The counselor arranges to see them again a month later.

At the next meeting, Anna is able to think about the options, but George is still not really ready. They decide that they will delay having a family for some time, and will talk to the counselor before Anna comes off the pill. Two years later, Anna and George return to the clinic. After further discussion about treatment of thalassemia, they decide to have a baby, without any form of prenatal testing. The baby is tested at birth for thalassemia. She is unaffected. In their second pregnancy they opt for a test, stating that they have realized since becoming parents how hard it would be to have a child who constantly needed additional care. They also don't feel they should 'try our luck twice'. The second child is also unaffected.

avoiding exposure to infection that may adversely affect the fetus such as listeria, cytomegalovirus and toxoplasmosis.

In addition to this general advice, a simple enquiry about whether either prospective parent has a family history of any genetic condition can enable specific concerns to be addressed before a pregnancy. In cases in which the health professional is unsure of the relevance of a family history, the staff of genetic services are generally very prepared to informally discuss the situation and advise whether referral to genetic services is warranted.

2. Maternal age

Maternal age can have an influence on the health of both the mother and the fetus. The older mother may be more prone to complications because of underlying medical conditions, for example, a mother in her 40s will be more likely to be affected by hypertension, varicose veins or stress incontinence than a woman in her 20s.

From the fetal perspective, chromosomal abnormalities are much more likely to occur in the fetus of an older mother [1]. The risk of all chromosome abnormalities increases with maternal age, because the chance of a nondisjunction event increases with the age of the oocytes (egg-producing cells).

In many areas amniocentesis is offered to women above a certain age. The actual maternal age above which amniocentesis was offered routinely varied according to local policy, but was usually 35, 36 or 37 years. With the advent of ultrasound scanning and maternal serum screening, this has altered. However, before starting a pregnancy, couples who seek pre-conceptual counseling should be aware of the increased risks to the fetus of a chromosomal abnormality, due to maternal age effect.

3. Neural tube defects and the role of folic acid in lowering the risk

The neural tube includes the brain and the spinal cord. This develops early in fetal life, and interruption to closure of the tube causes both spina bifida and anencephaly. Closure of the neural tube begins in the area of the cervical spine, with the spinal column forming distally in the direction to the coccyx, while the cranium closes over the brain. Failure of this process results in a neural tube defect (NTD). Although the appearance and effects of spina bifida and anencephaly differ, they are considered to be different manifestations of the same type of malformation, and are grouped under the term neural tube defect.

In the absence of other abnormalities in the fetus, neural tube defects are not inherited as a Mendelian disorder, that is, they do not follow one of the known patterns of inheritance. However, empirical data shows that couples who have one child with a NTD are at greater risk of having a second child with a NTD than other couples in the general population. Following the diagnosis of one child with a NTD, the risk of NTD in each subsequent pregnancy for that couple is ~ 4%. However, the risk can be lowered dramatically (to ~ 1%) by maternal folic acid supplements.

In 1991, a study of the effect of maternal supplements of folic acid on the rate of NTD in their offspring was reported [2]. The randomized control trial had to be discontinued before the end of the study, as it became very clear that folic acid was having an effect in reducing the number of children born with spina bifida. It was therefore unethical to continue to place women in the control group, as this was knowingly exposing the fetus of each of those women to greater risk.

Since that time, it has been recommended that women who may become pregnant take daily folic acid for at least two months before conception, and for the first 12 weeks of pregnancy. The dose for women whose risk of having a child with a NTD is considered to be at the population level is ~ 0.4 mg daily, but women whose children appear to be at higher risk are advised to take 4–5 mg per day. Whilst dietary advice is important, increasing the levels of folic acid in the maternal diet is not considered sufficient, as the rate of NTD does not appear to vary significantly with maternal diet. It may be that some women do not absorb folic acid from the diet as well as others.

Folic acid supplements are advised pre-conceptually because it takes some weeks to raise the maternal levels, and because the neural tube has already started to form before the mother is aware of her pregnancy. This means of course that some women will be taking folic acid for many months, even years, before conception occurs. However, as a B group vitamin, excess folic acid is excreted in the urine, and toxic levels do not therefore occur in the mother.

If a woman who has not been taking folic acid supplements becomes pregnant, she should start taking the supplements and continue until she is 12 weeks pregnant, although disruption to the normal formation of the neural tube may already have occurred. Folic acid tablets in the 0.4 mg dose are available in the UK and USA over the pharmacy counter, but the higher dose tablets are only provided on prescription. Prospective mothers who have had a previous child with a NTD or who are on anti-epileptic medication should be prescribed the higher dose because of their higher risk of having a child with a NTD.

KEY PRACTICE POINT

Women who may become pregnant should be advised to take folic acid for at least 2 months before conception and for the first 12 weeks of pregnancy.

4. Genetic conditions affecting the mother

4.1 Skeletal dysplasia

There are a large number of different types of skeletal dysplasia. This is a term used to describe a number of conditions in which the skeleton forms in an unusual way, causing bony deformity. The genetic code for the formation of the skeleton is faulty. A child may be at risk of skeletal dysplasia if either parent is affected, but if the mother is affected then pre-conceptual care is useful, as the condition of the pelvis and spine can be assessed prior to pregnancy. The bony deformity may affect the lie of the fetus in the third trimester, and may make passage of the fetus through the birth canal difficult in labour. As the fetus should not be exposed to X-rays, examination and investigation of the mother prior to pregnancy is important. The management of the pregnancy and delivery can then be planned. A similar situation may occur if the mother has a neural tube defect.

4.2 Connective tissue disorders (Ehlers Danlos syndrome and Marfan syndrome)

The management of pregnancy in women with a connective tissue disorder is discussed more fully in *Chapter 6*. There are two main issues concerned with the management of pregnancy, and it is useful to have discussed these before conception. First, the woman should have a baseline cardiac assessment if the condition is one in which cardiac complications can occur. Second, tissue friability should be assessed as the stretching of the uterine tissue in pregnancy may lead to uterine rupture, and if this is a significant risk the couple should be aware of this before pregnancy occurs.

4.3 Maternal diabetes

Maternal diabetes is known to affect the growth and development of the fetus. In poorly controlled diabetic mothers, there is an increase in the rate of spontaneous abortions. Intra-uterine growth retardation occurs more frequently in the babies of diabetic mothers, but the exact mechanism for this slowing of growth is not yet known. However, rather than being a direct effect of abnormal glucose levels, it is more likely to be due to a complex interaction between glucose levels, ketone bodies and decreased availability of insulin. Macrosomia of the fetus may also occur.

Congenital malformations occur significantly more frequently in infants of diabetic mothers, in fact the risk is between 2 and 4 times the population risk. The teratogenic effects seem to occur very early in the pregnancy (up to the seventh week). For this reason pre-conceptual counseling of diabetic mothers is helpful, as the most critical period occurs often before the mother is aware of the pregnancy. Strict blood control of blood glucose levels has been shown to decrease the rate of congenital malformations. In a regime designed to eliminate hyperglycemia, hypoglycemia may sometimes occur, but this is not thought to be detrimental to the fetus, and is preferable to hyperglycemia [3].

4.4 Epilepsy

Epilepsy is a condition which is thought to be multifactorial, that is, the offspring of a person with epilepsy has a greater chance than others in the general population of developing the condition. However, it is not inherited as a single gene disorder. If a mother has epilepsy, there is a danger of fetal hypoxia during a fit, and therefore the aim of treatment must be to avoid seizures during the pregnancy. The prospective mother should be advised to try and stabilize the epilepsy before conception, to reduce the possibility of fits. However, the use of drug therapy is not without additional risk to the fetus, as many anticonvulsants are known to cause serious congenital abnormalities.

The evidence appears to indicate that use of more than one drug to control the epilepsy increases the risk to the fetus. The aim therefore is to stabilize the mother prior to pregnancy, on the drug that is least harmful to the fetus, if possible. The woman with epilepsy should always be referred to a physician experienced in care of patients with epilepsy before pregnancy for this purpose. Women who are taking anticonvulsants are in the group considered to be at increased risk of having a child with a NTD and should therefore be prescribed the higher dose folic acid (5 mg daily), to be taken pre-conceptually for at least 2 months and for the first 3 months of pregnancy [4–6].

CASE EXAMPLE **JODI**

Jodi is a 27-year-old woman who is married to Steven. They have been together from their teens, and plan to have a family. Jodi has been epileptic since the age of 7 years, and is on carbamezipine. Controlling her epilepsy has been very difficult but this drug seems to be effective.

Jodi asks her GP about the risks of her epilepsy being passed on to her children, and he refers her to the genetic counselor in the local genetics service. The counselor tells her the risk of her children having epilepsy is low, Jodi confides that she is really worried about her child being disabled. The counselor discusses the risks of congenital abnormalities due to the anticonvulsants. These include clefting, spina bifida and learning difficulties. Jodi is referred back to her physician for review of her medication; alteration to her drug therapy is not possible as most other anticonvulsants have been tried in her case.

Jodi and Steven decide to go ahead and she becomes pregnant several months later. A detailed scan is ordered at 18 weeks gestation, and the fetus is found to have a neural tube defect. Steven and Jodi have already decided that they cannot raise a child with physical disabilities, and terminate the pregnancy.

---

5. Maternal drug therapy

It is not possible to list the drugs with potentially harmful effects on the fetus. Each case needs to be treated individually, and an assessment made of the benefits to the mother and fetus as opposed to the possible adverse effects. Obviously drug therapy that is not strictly necessary to maintain the health of the mother should be avoided in pregnancy. Details of the potential effects of drugs and possible alternative drugs can be obtained through a pharmacist. In some situations, alternative therapies such as aromatherapy or acupressure may be helpful.

6. Consanguinity

Pre-conceptual advice may be sought by couples who are related biologically (consanguineous couples). There appears to be a great deal of traditional superstition attached to cousin marriage in some communities, making the couple apprehensive about the genetic advice they may receive. In other ethnic groups, cousin marriage is the norm.

In fact, consanguineous couples are only slightly more likely to have a child with a genetic condition than other unrelated couples. They are at increased risk of having a child with a recessive condition, as both parents could carry the same faulty gene, inherited from the common grandparent or great grandparent (*Figure 1*). All couples have a 2–3% chance of having a child with a serious health concern, and couples who are first cousins have a risk of ~ 5%. The children of second cousins (or those who are more distantly related) will be at less risk of a recessive condition. If there is any evidence of a recessive genetic condition in the family, carrier testing for both partners may be indicated [7].

In the UK there is no legal barrier to marrying a cousin. Marriage between closer relatives such as between siblings, parent/child, grandparent/grandchild, uncle/niece or aunt/nephew is not permitted. In some states in the USA, first cousin marriage is illegal.

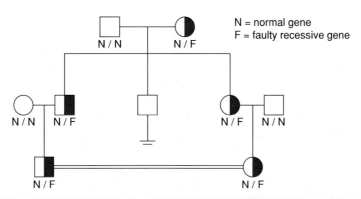

Figure 1. Potential inheritance of a faulty gene from a common grandparent.

7. Atopy and prenatal exposure to food allergens

Allergy to certain foodstuffs can produce a severe anaphylactic reaction. The most well-known example of this type of allergen is peanut, and death may occur in sensitized individuals as a result of ingesting even minute quantities of peanut or peanut oil in food. There is evidence that in families with a history of allergy, infants may be sensitized to certain allergens while still *in utero* [8]. Advice from the Department of Health in England now states that women who have a history of eczema, asthma, hay fever or other allergies, or who have close family members with these conditions, should avoid foods such as peanuts during pregnancy. This advice is best given before the woman is pregnant, so she can limit the content of these foods in her diet while trying to conceive and in early pregnancy, before the pregnancy is confirmed.

8. Preparing for prenatal diagnosis

One important aspect of pre-conceptual care relates to families where there is a known genetic risk. Although a great number of genetic conditions occur sporadically, in many cases there is a family history of a condition. If a couple are concerned about a history of a disorder, whether this has occurred in a previous child of theirs or in members of the family, they should be referred to genetic services for assessment and information.

Where the couple would wish to have a prenatal diagnostic testing on the fetus, preparatory work needs to be undertaken. The laboratory will often require samples from family members prior to the testing, to enable the pattern of genetic markers in the family to be ascertained, or to enable the relevant gene mutation to be identified. It is not uncommon for this work to take at least several months. In many situations in genetics, it is only after this preliminary work is done that the family can be offered the opportunity to have prenatal diagnosis with certainty. Of equal or greater importance is the preparatory emotional work and decision-making that is preferably done by the couple before the pregnancy. The genetics team will discuss the options with the family, and support them in making the decisions about testing that are best for them in their unique circumstances. If the couple want to consider prenatal diagnosis, they should be advised to contact their midwife or doctor as soon as the pregnancy is confirmed, so that arrangements for early scanning (to assess gestation accurately) and testing can be made. When testing for rare disorders, the laboratory needs to be aware that a test is planned so that staff and materials can be made ready.

KEY PRACTICE POINT

Any couple that is concerned about a genetic risk to their future children should be referred to the genetic service *before* pregnancy whenever possible so preparations can be made for prenatal diagnostic tests and so the couple can discuss their options.

CASE EXAMPLE   SUZANNE AND HELEN

Suzanne and Helen are sisters. When Helen became pregnant for the first time, her midwife asked her about a family history of any genetic condition, and Helen told her that her younger brother had muscular dystrophy. The midwife referred her for genetic counseling. Helen was found to be a carrier of Duchenne muscular dystrophy, an **X-linked** genetic condition that causes death in the late teens. Helen and her husband decided to have a prenatal test, and the male fetus was found to have inherited the condition. Helen and her husband reluctantly decided to terminate the pregnancy. This was very difficult as they had been overjoyed at the confirmation of the pregnancy, with no real knowledge about the risk to the baby. Helen's sister Suzanne asked to see the genetic nurse. She had been attending an infertility clinic for 3 years, and was currently receiving clomiphene therapy to help her conceive. She had no idea she might be a carrier of the condition that affected her brother. She was also found to be a carrier of Duchenne muscular dystrophy. She discontinued therapy, as clomiphene greatly increases the risk of multiple pregnancy and she could not face the thought of terminating several babies or of selective feticide. Suzanne felt angry that the risk of muscular dystrophy had never been discussed with her prior to her infertility treatment.

9. Issues of fertility

There are a number of genetic conditions which are known to cause subfertility or infertility. These include conditions related to imbalance of the sex chromosomes such as Turner syndrome in the female (45,X), and Klinefelter syndrome in the male (47,XXY). Female carriers of Fragile X may experience premature ovarian failure, and males with cystic fibrosis may have oligospermia or complete absence of the vas deferens. New reproductive technologies may be able to assist these clients to have a child, although in many cases this is still not possible. It is always worthwhile to offer such couples a referral to an **assisted reproduction** unit for assessment, whilst not raising inappropriate expectations.

The use of assisted reproduction techniques has raised new issues for infertile men, as it is now thought that male infertility is often the result of a gene mutation on the Y chromosome. Previously men who carried this mutation would not have had the chance of passing it on to their offspring, but with the new methods they are able to have children and thus produce sons who are similarly infertile [9].

If a mother presents having already had several miscarriages, a balanced chromosome translocation in one of the parents should be suspected (see *Chapter 4*). Chromosome studies for both parents should be offered.

10. Adoption

There are two main aspects of adoption that require some consideration in relation to genetics. First, the rights of both the child and the prospective parents when a child with a genetic condition is offered for adoption. It is desirable that the parents be given full and appropriate information about the effects of the condition and the prognosis for the child, in order to make adequate provision for the care of that child, although there are conflicting views about pre-adoption genetic testing [10]. It is our experience that adoptive parents of a severely disabled child approach the child's condition differently to natural parents, perhaps because natural parents may feel an element of guilt, which is not present in adoptive parents.

Second, there are issues that may arise if a parent who is affected by or at risk of a genetic condition wishes to adopt a child. It is the responsibility of those arranging the adoption to make an assessment in the best interests of the child, and therefore these genetic factors may be taken into account. For example, if a prospective parent is at risk of a condition that might seriously affect their ability to parent, the adoption agency may not approve the adoption. This can be very difficult for couples who wish to avoid having a child at risk of the condition, and who then feel they are at a disadvantage in terms of the possibility of adoption as well.

11. Conclusion

This chapter has addressed the issues that might arise prior to conception, and has emphasized the need for pre-conceptual care in a variety of situations. There is much that can be done pre-conceptually to help a couple achieve a healthy pregnancy, and planning for prenatal genetic testing is much better done before the woman is pregnant. The next chapter focuses on the genetic issues of relevance during pregnancy.

TEST YOURSELF

Q1. A woman attending a family planning clinic says she is planning a pregnancy in the next few months. What should you as a health professional consider discussing with her?

References

1. Polani PE, Alberman E, Berry AC, Blunt S, Singer JD (1976) Chromosome abnormalities and maternal age. *Lancet* **1** (7984): 516–517.
2. MRC Vitamin Study Research Group (1991) Prevention of neural tube defects: results of the Medical Research Council vitamin study. *Lancet* **338** (8760): 131–137.
3. Eriksson U (1997) Embryo development in early pregnancy. In A Dornhurst and D Hadden, editors. *Diabetes and Pregnancy*. Chichester: Wiley & Sons.
4. Lewis DP, Van Dyke DC, Stumbo PJ, Berg MJ (1998) Drug and environmental factors associated with adverse pregnancy outcomes. Part I: antiepileptic drugs, contraceptives, smoking, and folate. *Ann Pharmacother* **32** (7–8): 802–817.
5. Samren EB, van Duijn CM, Koch S *et al.* (1997) Maternal use of antiepileptic drugs and the risk of major congenital malformations: a joint European prospective study of human teratogenesis associated with maternal epilepsy. *Epilepsia* **38** (9): 981–990.
6. Hill RM (1973) Editorial: teratogenesis and antiepileptic drugs. *NEJM* **289** (20): 1089–1090.
7. Harper PS (1998) *Practical Genetic Counselling*, 5th edn. Oxford: Butterworth-Heinemann.
8. Hourihane JO, Dean TP, Warner JO (1996) Peanut allergy in relation to heredity, maternal diet, and other atopic diseases: results of a questionnaire survey, skin prick testing, and food challenges. *BMJ* **313** (7056): 518–521.
9. Simpson JL, Lamb DJ (2001) Genetic effects of intracytoplasmic sperm injection. *Semin Reprod Med* **19** (3): 239–249.
10. Jansen LA, Ross LF (2001) The ethics of preadoption genetic testing. *Am J Med Genet* **104** (3): 214–220.

Further reading

Harper PS (1998) *Practical Genetic Counselling*. Oxford: Butterworth-Heinemann. (General reference book for all health professionals.)

6 Pregnancy and the perinatal period

1. Introduction

Genetic counseling is often inextricably linked in the minds of health professionals with reproduction. Whilst it is hoped that this book will help to illustrate that genetic issues are of relevance to people in all stages of life, it is clear that many families with concerns about a genetic condition will seek information during pregnancy and in the weeks after birth. Of course, a significant number of families will become aware of their genetic risk for the first time during that period, as a result of prenatal testing, ultrasound scanning or congenital abnormalities diagnosed in a newborn. In this chapter, we describe common situations in which genetic conditions may become apparent, the types of tests available and some guidelines for good practice.

2. Care of the mother during pregnancy

When booking a client for antenatal care, the midwife has a responsibility to ascertain whether the mother or fetus is at any particular genetic risk. At the initial maternity consultation, questions such as those listed in the UK National Maternity Record (NMR) should be used to ascertain whether there may be an increased risk of a genetic condition in the family. This record has recently been altered to facilitate taking a genetic history and provides guidelines for action by the midwife should any concerns be raised. If a version of the NMR is not used, these questions can still be used to guide the midwife in taking a relevant history.

If the mother has a genetic condition, this may alter the management of the pregnancy. Because of improved health care, women with serious inherited conditions may now be sufficiently healthy to enable them to conceive and carry a pregnancy to term. These mothers should always be jointly managed by an obstetric team experienced in the care of woman with high-risk pregnancies and a physician experienced in the care of patients with the particular condition [1,2].

Specific examples are given below.

CHECKLIST OF GENETIC CONDITIONS TO BE USED BY THE MIDWIFE WHEN BOOKING A WOMAN FOR ANTENATAL CARE

Questions related to the mother's care:

Diabetes	No/Yes
Blood clotting problems (DVT, PE)	No/Yes
Stillbirth/multiple miscarriages	No/Yes

If considered significant, referral to a consultant obstetrician should be considered.

Questions related to a risk to the fetus:

Has anyone in either your own family or your partner at the time of conception's family:

	Your family	Partner's family
A disease that runs in families?	Yes/No	Yes/No
Been seen by a geneticist or genetic counselor	Yes/No	Yes/No
Had learning difficulties (e.g. needed special help at school)	Yes/No	Yes/No
Had a baby with abnormalities present at birth	Yes/No	Yes/No
Had more than two stillbirths or miscarriages	Yes/No	Yes/No

If considered significant, referral to an obstetrician, fetal medicine or clinical genetics specialist should be considered:

Serum screening	Offered: Yes/No	Accepted: Yes/No (result)
Ultrasound scan	Offered: Yes/No	Accepted: Yes/No (result)

Sickle cell and/or thalassemia carrier screening offered/accepted

N.B. Questionnaire developed by the UK Joint Committee for Medical Genetics

---

2.1 *Marfan syndrome*

Marfan syndrome is a dominant condition that affects a number of different body systems. The genetic defect adversely affects the development of connective tissue, namely collagen. Marfan syndrome causes tall stature, hypermobility of the joints, cardiovascular abnormalities, aortic root dilatation and aortic aneurysm, and dislocated lenses. Owing to the joint laxity and stature, serious back pain in pregnancy may become a problem. The mother with Marfan syndrome who has an existing cardiovascular abnormality should, of course, be under the care of a cardiologist, who should be made aware of the pregnancy. However, if Marfan syndrome is suspected or diagnosed, and the mother is not under the care of a cardiologist, then an urgent referral should be made for cardiac assessment. Sudden death from cardiac complications may occur in untreated cases [3–5].

2.2 Ehlers–Danlos syndrome

The term Ehlers–Danlos syndrome (EDS) refers to a group of disorders of connective tissue. Whilst hyperextensibility of skin, increased mobility of joints and tissue fragility are common features, the severity of the condition varies widely. Women with this condition may be at higher risk of spontaneous abortion and stillbirth. It is thought that collagen in the chorionic membranes is affected by the condition, making premature rupture of membranes more likely. Because of the fragility of vessel walls, these women will also be at increased risk of hemorrhage antenatally and during labor. The joint laxity experienced by these women may result in dislocation of the symphysis pubis, whereas the frailty of other tissues makes varicose veins and abdominal herniae more common in this group of patients [6]. Additional monitoring of the mother during labor is indicated because of the risk of uterine rupture. Wound dehiscence may occur after either cesarean section or vaginal delivery, and the use of supportive aids such as tapes or nonabsorbable sutures is advised.

2.3 Maternal phenylketonuria

Phenylketonuria (PKU) is a recessive condition, in which the biochemical pathway involved in the metabolism of phenylalanine is disturbed, causing increased phenylalanine levels in the affected person. The accumulation of phenylalanine in the body results in mental retardation, however, the use of a low phenylalanine diet during infancy, childhood and adolescence will dramatically reduce the risk of mental retardation.

It has been shown that during pregnancy high maternal levels of phenylalanine have adverse effects on the fetus, causing microcephaly, mental retardation, dysmorphic features and congenital heart disease. Although women with PKU who are planning a pregnancy can revert to the diet some months before conceiving, thus reducing serum phenylalanine levels, in an unplanned pregnancy the fetus would already be exposed to high phenylalanine levels before the pregnancy is diagnosed. It is therefore strongly advised that women are encouraged to remain on a low phenylalanine diet, to maintain a serum phenylalanine level of 120–360 µmol/l until they have completed their families, to reduce the risk to their offspring [7,8].

2.4 Cystic fibrosis

Cystic fibrosis (CF) is a common recessive condition with approximately 1 in 20 people in the UK and USA populations being carriers. The gene mutation causes a disturbance of chloride, sodium and water ratios in secretions, resulting primarily in respiratory infections, reduced lung function, and reduced pancreatic function. Improved management of CF has enabled some affected women to live into adulthood and maintain sufficiently good health to consider having a family.

Where the mother's lung function is normal, the CF does not appear to confer significantly higher risks to mother or fetus, however, where there is impaired function there is a greater risk of prematurity for the fetus, and therapeutic abortion may be considered to preserve the mother's health. Mothers with CF are prone to develop diabetes and this should be borne in mind when caring for such women. It is important that the mother should be under the care of a respiratory specialist for monitoring at least monthly during the pregnancy [9].

2.5 *Hemoglobinopathy*

Thalassemia major. In most centers caring for women with thalassemia, a protocol will be used for treatment with a series of blood transfusions. Iron chelation therapy (desferrioxamine) will also usually be administered to maintain stable serum ferritin levels. Cardiac problems, diabetes and hepatitis C may additionally complicate the pregnancies of women with thalassemia major [10]. Cardiac, hepatic and endocrine function should be monitored during pregnancy and in the postpartum period [11]. Delivery by cesarean section may be planned prior to the due date, usually at between 37 and 38 weeks gestation [12].

Sickle cell disease. Mothers with sickle cell disease are at increased risk of spontaneous abortion and stillbirth [13]. Severe anemia in pregnancy (hemoglobin <60–70% normal) is treated with blood transfusion. Trials of prophylactic transfusions to improve outcomes of pregnancy have not been conclusive [14].

2.6 *Congenital heart disease*

Congenital heart disease due to a genetic condition may not be treated any differently than that which occurs sporadically, however, there may be additional health factors to be considered in women who have a genetic condition affecting organs outside the cardiovascular system. For example, a woman with heart disease due to a chromosome 22q microdeletion may have learning difficulties and problems with immunity.

Prophylactic antibiotics are often considered during labor. Whilst vaginal delivery is recommended for many conditions, a Cesarean section may be the best option in a woman who is already cyanosed and who is being delivered prematurely, and for those women with Marfan syndrome. A detailed article on the treatment of women with specific heart defects has been written by Oakley [15] and we would commend that to any midwife seeking guidance about the care of a particular woman.

In some cases, a congenital heart defect may be so severe that pregnancy is contraindicated and termination of pregnancy to preserve the life of the mother may be advised by a cardiologist [1].

3. Spontaneous abortion (miscarriage)

Spontaneous abortion is thought to occur in ~10–15% of confirmed pregnancies [16]. Unfortunately, women with a history of miscarriage are less likely to carry subsequent pregnancies to term. There are many causes of spontaneous abortion, these include:

- anatomical abnormality (e.g. bicornuate uterus);
- infection (e.g. herpes, cytomegalovirus or rubella);
- maternal hormonal imbalance (e.g. progesterone deficiency);
- endocrine dysfunction in the mother (e.g. poorly controlled diabetes);
- maternal immunological abnormality (e.g. systemic lupus erythematosis).

However, an unbalanced chromosome arrangement in the fetus is believed to be the cause of up to 50% of spontaneous abortions [17]. This explains why the percentage of pregnancies that spontaneously abort increases with maternal age [18].

In any pregnancy, there is a risk that the fetus will have an unbalanced chromosome arrangement, but many of these pregnancies result in spontaneous abortion even before the mother is aware that she has conceived. The chromosome imbalance may occur sporadically, or more infrequently it is inherited from a parent with a chromosome translocation.

Because most cases of chromosomal imbalance in a fetus occur sporadically, the parents have a low risk of the same problem occurring for a second time. Thus, a mother who has conceived one child with a chromosomal abnormality will usually have a < 1% chance of having another child with that condition [19].

A balanced translocation as a cause of spontaneous abortion is usually not suspected until there have been at least two, but usually three, pregnancies lost in this way. After recurrent abortion, it is good practice to investigate the chromosome patterns of both parents [20]. If one of the parents has a balanced translocation or rearrangement, there will normally be at least a 50% chance that in each pregnancy the fetus will inherit a balanced chromosome arrangement and therefore will develop normally. In each individual case, information should be sought from the cytogeneticist or genetic counselor about the risks to the particular couple, and whether prenatal diagnosis is advised in future pregnancies.

Although many parents find losing a pregnancy very distressing, some are able to rationalize it by concluding that the fetus was not developing normally and that the abortion was 'nature's way' of preventing the birth of an abnormal child. Whilst this thought may not lessen their sense of loss, it may help them to make sense of what has happened. Many parents express their sense of frustration and hurt that their friends or family do not regard the expected baby as 'real' and therefore minimize the loss, or ignore it altogether. The grief accompanying the loss of a baby has been thoughtfully dealt with by Kim Kluger-Bell in her book *Unspeakable Losses*, which we would recommend to those who care for couples during or after pregnancy (see Further reading).

KEY PRACTICE POINT

After a couple have experienced a spontaneous abortion, explore with them the meaning of the miscarriage to them. For some, it will be 'just one of those things', for others a deeply felt loss. If they have named the baby, use the given name and acknowledge the reality of the loss.

CASE STUDY **TERRY AND ROSE**

Terry and Rose were referred to the genetic service after experiencing three miscarriages. Their family doctor had taken a blood sample for karyotyping after the loss of the third pregnancy, and Terry was found to have a balanced translocation of chromosomes 2 and 6.

At first, the couple were pleased to have discovered the cause of their failure to carry a pregnancy to term, and they were encouraged when told there was a 50% chance of success in each pregnancy.

Terry was one of four children. His brothers and sister were offered chromosome analysis, as were his parents. Terry's mother had four normal pregnancies with no spontaneous abortions, so it seemed unlikely either of his parents carried the translocation.

However, Terry's father was found to carry the same translocation. Although Terry and Rose acknowledged that this meant they were likely to have a live baby, they also felt very angry that they had been so unlucky in losing three pregnancies, when Terry's parents had lost none. One of Terry's brothers was also found to carry the translocation.

Rose became pregnant again, and aborted at 6 weeks gestation. The couple then started to consider artificial insemination using donor sperm, as they felt that each miscarriage was devastating and they could not bear any more emotional strain. Whilst awaiting an appointment for insemination, Rose again became pregnant, only to miscarry a fifth pregnancy.

At this stage, the risk figure of 50% seemed inaccurate to them, how could they lose five pregnancies in a row?

A letter came from the reproductive medicine clinic, stating they would need to discuss the options fully before treatment was offered. Both partners were very upset, feeling this cast doubts on their ability to decide the best course of action for themselves and minimizing the distress they had suffered. They cancelled the appointment.

When Rose became pregnant for the sixth time they hardly dared to hope. However, this time the pregnancy proceeded normally. They discussed prenatal diagnosis, but decided against an invasive procedure as they did not want to take any risk of losing a normal pregnancy due to complications of amniocentesis. They opted for high resolution ultrasound but were adamant that even if an abnormality was found they would not wish to terminate the pregnancy.

Terry and Rose had a normal baby girl at term. They have decided not to try and increase the family as they are too fearful of the pain of another miscarriage.

4. Down syndrome (trisomy 21)

Down syndrome occurs when a fetus has developed with an additional copy of chromosome 21. This usually occurs sporadically, because the ovum or sperm has been formed with two copies of chromosome 21, rather than the normal single copy. The recurrence risk is therefore usually low. Before advising parents, it is important to check the report of the chromosomal structure of the affected child. If the result shows that the child inherited a full additional copy of chromosome 21 (i.e. 47, XX, +21 or 47, XY, +21), then it is highly likely to have occurred sporadically. In this case, the risk for future pregnancies for this couple will be 1% or twice the mother's age-related risk, whichever is higher. If, however, the chromosome result shows the baby with Down syndrome had a translocation, the parents' chromosomes should be studied to ascertain whether one of them carries a Robertsonian translocation, as this would increase their chances of having a second child with a chromosomal abnormality [21] (*Figure 1*).

Age of mother at time of baby's birth (years)	Risk of baby affected with Down syndrome
15	1 in 1580
20	1 in 1530
25	1 in 1350
30	1 in 901
35	1 in 385
37	1 in 240
39	1 in 145
41	1 in 85
43	1 in 49
45	1 in 28
47	1 in 15
49	1 in 8

Figure 1. Maternal age related risk table for Down syndrome (live births)[21]

5. Prenatal testing techniques

All healthcare professionals who come into contact with a pregnant couple
should be aware of the types of prenatal tests that are possible. Screening test
is the term used for any test in which the result provides an indication of the
level of risk in a particular pregnancy; this category includes maternal serum
screening and nuchal translucency measurement under ultrasound. A
diagnostic test (such as fetal karyotype after amniocentesis or chorionic villus
sampling; CVS) usually provides the parents with a definite result. The
majority of screening tests and diagnostic procedures are carried out in
women who are not at increased risk and in whom any abnormality found
would not have been predicted. The midwife has a particular responsibility
when discussing choices for routine antenatal diagnosis to insure that the
woman is given enough information to allow her to make a truly informed
choice. Sadly, in the past, both medical and nonmedical practitioners have not
been shown to have done this [22]. This issue is going to become even more
relevant as screening for abnormality becomes increasingly part of routine
antenatal care. It is important to recognize that at the present time the
outcome of most prenatal diagnostic tests is knowledge that can lead to a
decision whether to terminate a pregnancy. Although this is a valid choice, it
may be a choice that some couples do not want to have to make. If antenatal
screening becomes routine then it is essential that the full consequences of
screening be explored with a pregnant couple in order that they can make the
choice that best fits their own beliefs, values and particular circumstances.

In any situation, the offer of a screening or diagnostic test should not be
conditional on the parents agreeing to termination if the fetus is found to be
abnormal. That decision remains with the parents, and some may choose to
continue with a pregnancy. However, they should be aware that some tests are
invasive, and as such carry a risk to the fetus. If the results of the test would
not affect their continuing with the pregnancy, taking that risk may not be
justified.

Couples that choose to continue with a pregnancy when the fetus has been
diagnosed as having abnormalities or health problems require specific support
from the midwifery team. Delivery may be planned in a center where access
to surgical teams or specialist care is available. Routine care may be altered

during labor, as fetal monitoring might not be appropriate, and the method of delivery should also be planned, as Cesarean section delivery for a mother whose child will not survive would probably not be in the best interests of the mother.

If a couple choose to terminate a pregnancy, this is usually done via suction evacuation of the uterus under general anesthetic before 12–14 weeks' gestation. After that time, labor is induced, often by the use of prostaglandin pessaries and intravenous oxytocin. As the cervix is usually firm, labor often lasts many hours, even several days. Sensitive support for both parents and appropriate analgesia for the mother are imperative.

5.1 *Antenatal screening for Down syndrome*

Antenatal screening programs in the UK currently focus mainly on detection of Down syndrome. It is known that increasing maternal age leads to a higher risk of having a baby with Down syndrome [23]. Amniocentesis is usually offered to all women who will be over a certain age at the birth of the baby, normally 35 or 37, dependent on the policy of individual health districts or institutions. However, even though an individual's risk of having a baby with Down syndrome is low, the majority of pregnancies happen in women under the age of 35, and therefore most babies with Down syndrome are born to mothers aged less than 35 years. Various screening methods [24] involving biochemical testing of the mother's blood and ultrasound scanning have been developed in an attempt to identify individual women whose pregnancy may be at increased risk. These include:

First trimester:
- nuchal translucency measurement by ultrasound;
- double test on maternal serum, pregnancy-associated plasma protein A (PAPP-A) and human chorionic gonadotrophin (HCG);
- combined nuchal translucency and double test.

Second trimester:
- double test on maternal serum, alpha fetoprotein (AFP) and HCG;
- triple test on maternal serum, AFP, HCG and unconjugated estriol (uE3);
- quadruple test on maternal serum, FP, HCG, uE3, inhibin A.

The *integrated test* involves testing in both first and second trimester:
- nuchal translucency and PAPP-A in first trimester;
- quadruple test in second trimester.

However, these tests only provide an estimation of the chance the fetus will have Down syndrome, rather than a definitive result. If the risk is > 1 in 250 (this threshold may vary according to the policy in the maternity unit), the mother is usually offered a test to check the karyotype of the fetus. Although the purpose of screening is to detect Down syndrome, the biochemical markers studied in maternal serum may be altered by other fetal abnormalities and the diagnosis of a completely unexpected abnormality may be made [25]. The serum screening test result provides the parents with more information on which to base decisions about definitive testing, but the majority of those who have a serum screen result in the 'high-risk' category will have a normal fetal karyotype (false-positive serum screen result). It must also be remembered that a serum screen result in the low-risk category does not rule out the possibility that the fetus does have Down syndrome (false-negative serum screen result). Increasingly, women are being offered nuchal translucency measurement as an alternative or addition to serum screening. A measurement is made of the fluid filled space at the back of the neck, at the end of the first trimester. An unusually large measurement may indicate a fetus with Down syndrome or other abnormality.

New method

New techniques for extracting fetal cells from a maternal blood sample are being developed. When successful, these techniques will give couples the option of knowing the definitive karyotype of their expected baby without the risk of invasive tests.

5.2 Ultrasound scanning

Ultrasonography is now a routine part of most women's antenatal care. It should be emphasized to the mother that one of the purposes of scanning is to detect abnormality rather than to provide reassurance of a normal pregnancy, although the outcome is normal in most routine scanning. The sensitivity of the ultrasound scan at detecting abnormality will depend on the stage of pregnancy and the nature of the abnormality. For example, studies have shown that detailed scans in the second trimester detect between 88 and 100% of central nervous system abnormalities, but only between 36 and 66% of major heart abnormalities [26,27]. If the pregnancy is at a higher risk of a heart abnormality detailed specialist cardiac scanning can be offered at a specialist unit. A family history of a genetic disorder or congenital abnormality may indicate the need for a higher level of scanning than that which is routinely offered. Early scanning can be useful for dating purposes and may detect gross abnormalities such as anencephaly, however, the early dating scan is not a substitute for a later anomaly scan.

Scanning may detect abnormalities that are difficult to interpret, and further more invasive testing may be required. Certain structural markers such as specific congenital abnormalities (e.g. heart defect, renal abnormality or skeletal abnormality) may be an indication for chromosome analysis of the baby.

5.3 Amniocentesis

Amniocentesis (*Figure 2*) is usually performed at around 16 weeks' gestation. It is normally offered to women who have an increased risk of a baby with a chromosome abnormality. This may be because of the mother's age or because other screening tests or the scan may suggest the pregnancy is at some risk. The amniotic fluid contains cells that are sloughed off the baby's

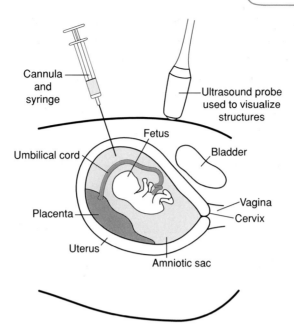

Figure 2. Aminocentesis.

skin, urinary tract, lungs, etc. These cells can be cultured and a karyotype performed. In a pregnancy that may be at risk of rhesus disease analysis of the bilirubin content of the amniotic fluid is used to assess the status of the baby. Amniocentesis can also be used to withdraw amniotic fluid for biochemical testing (for some specific metabolic conditions) or for viral studies if fetal infection is suspected.

The test is done under ultrasound guidance and a full bladder is necessary. Local anesthetic is injected into the skin of the abdomen and a fine needle is passed through the skin into the uterus and the amniotic cavity. A small amount of fluid is withdrawn into a syringe and the needle removed. The test is normally performed in the outpatient clinic (ambulatory medicine).

The time taken for the results will depend on local circumstances, but is normally at least 7 days. The result is a full karyotype, which will detect other abnormalities of chromosome number and structure in addition to Down syndrome. The implications of these other abnormalities may not be clear or may require expert interpretation. Some of them, such as abnormalities of the sex chromosomes may not be as significant as Down syndrome. Amniocentesis has a risk of causing a miscarriage of ~ 0.5% [28].

5.4 Chorionic villus sampling or biopsy

CVS is used to obtain a sample of tissue that can be used for karyotyping (*Figure 3*). However, DNA can be extracted from the chorionic villi more

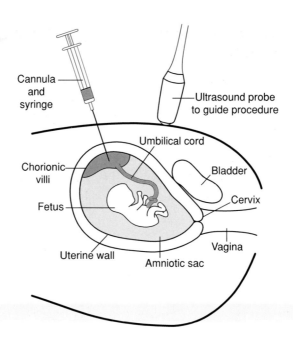

Figure 3. Chronic villus sampling.

easily than from the fetal skin cells, and therefore CVS is the method of choice for obtaining a sample for DNA analysis of the fetus. The chorionic villi are the part of the placenta that attaches into the wall of the uterus. Because the placenta and the fetus arise from the same fertilized embryo they essentially share the same genetic and chromosomal material.

The main advantage of CVS is that it is performed at about 10–12 weeks' gestation. The disadvantage is that it probably has a higher rate of miscarriage than amniocentesis. The exact miscarriage risk is difficult to determine as miscarriage is relatively common in early pregnancy. Individual centers usually maintain their own follow-up and will be able to quote a risk for their own center, which is normally in the region of 2–3% [29]. There has also been a concern about a risk of limb abnormality with CVS. It is thought currently that this is related to the timing of the CVS and for this reason the test is not performed earlier than 10 weeks' gestation [30]. An association between talipes and CVS has also been suggested [31].

CVS is normally performed through the abdomen in the same way as amniocentesis. It can also be done by inserting a fine cannula into the uterus, through the vagina and cervix. The result will normally take about a week. A result can be generated in 24–48 hours without culturing the cells. This analysis will not be at the same level of detail as a result from a culture and is more prone to error [32].

CHANGING OPTIONS IN PRENATAL DIAGNOSTIC TECHNIQUES

Using a family affected by Duchenne muscular dystrophy (DMD) as an example, we can trace the changes in genetic testing that can offer greater certainty to a family. Consider the options of a woman called Maria, who is a carrier of DMD.

1955: Maria is told that in any pregnancy she conceives, the fetus has a 25% chance of inheriting muscular dystrophy. There are no tests that can be done to clarify the status of the fetus during pregnancy. Maria's only options are to have a pregnancy and take the risk or avoid having any children.

1970: Amniocentesis for fetal sexing is now available. Maria is offered amniocentesis at 16 weeks' gestation, to determine the sex of her unborn child. To avoid having a boy affected by DMD, she will have to terminate all male pregnancies, even though each boy has only a 50% chance of having the condition.

1985: DNA testing of the fetus is now available, but blood samples are required from a number of affected and unaffected relatives. First Maria is offered prenatal diagnosis by 'linkage' studies, which track the pattern of X chromosomes through her family. If a male fetus is found to carry the high-risk X chromosome, there is a 90% chance that he will develop DMD.

2000: Definitive testing is possible. The deletion of the dystrophin gene on the X chromosome has been identified in Maria's family. After CVS, a male fetus can be tested to show whether he has inherited the deletion. If so, he will almost certainly develop DMD.

Although testing techniques have improved, and options are undoubtedly greater, the decision to terminate a pregnancy is generally difficult in any family. However, the family may be more certain that their decision was right for them when the outcome of the pregnancy is known with greater accuracy.

Harding Family

Referral letter

Dear

RE: Sarah Harding
This delightful 22-year-old patient of mine is unexpectedly pregnant. Her father was diagnosed with Huntington's disease 4 years ago. Sarah is aware that a test for this is possible but has always said she would not want to know herself. Obviously the pregnancy may change the picture. Please see and advise.

Sarah was offered an urgent appointment with the genetic team and an ultrasound dating scan arranged. Sarah came to the appointment accompanied by her older sister Lauren, and Sarah's partner Nick. The genetic counselor realized that she had met Lauren in the past when the diagnosis in their father had been made. At that time Sarah was doing her A-levels and had not wanted to be seen in the genetics department. The family file was reviewed and it became clear that the family had been known to the genetics department for many years. Various different members had been seen and in the past DNA samples had been collected for linkage studies. (See Chapter 4, prior to identification of the gene mutation, linkage studies allowed some people to have predictive tests.)

The genetic counselor asked Sarah about her family tree (Figure 4) in order to allow her to talk about what Huntington's disease meant to her. Sarah was well informed and was very sure that she did not want to know if she had the gene; she felt she would not be able to cope with living with the knowledge that she was going to develop Huntington's. She and Nick were also sure that they did not want to pass the gene on to the current pregnancy which they had already decided that they wanted to continue if possible. The dating scan had shown that Sarah was 8 weeks' pregnant.

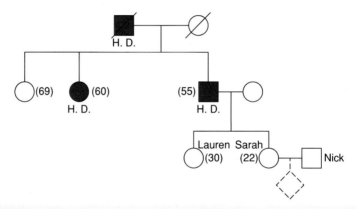

Figure 4. Family tree.

The counselor discussed the options that were available to Sarah and Nick:

(i) have a termination of the pregnancy on the basis it was unplanned;
(ii) continue with the pregnancy on the basis it was at 25% risk;
(iii) have a predictive test for Sarah and then test the pregnancy if she should prove to have the gene;
(iv) have an **exclusion test** on the pregnancy; this was possible because the family was already known to the department and some linkage analysis had already been done [33].

Sarah did not want to take the risk of continuing without testing, as she said she didn't wish to 'Put another person through what I'm going through now'. Nor did she want to know at this time if she had the gene mutation. Gene testing would have been a very rushed procedure because of the need to get the results back in order to do a CVS at 10 weeks if the predictive test on Sarah showed that she had the gene.

Sarah decided on an exclusion test using linked markers. Linkage results using markers closely linked to the Huntington's gene had shown that her father had inherited different genetic markers from his parents. In the past Sarah's mother had also given a DNA sample and the results had shown that she had different genetic markers at the Huntington's disease locus from her husband. This means that in Sarah it was possible to distinguish between the maternal and paternal markers (Figure 5).

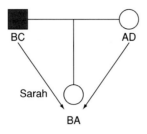

Figure 5. Linkage results.

In Sarah the marker A inherited from her mother would not be linked to the Huntington's disease gene. If this marker was passed on to the pregnancy it would be at low risk. In her father it is not known if the marker B or C is linked to the Huntington's disease gene. (Remember he will have one normal gene and one Huntington's disease gene.) However, because marker B is inherited from her affected parent, if this marker is passed on to the pregnancy it will be at the same risk as Sarah – 50%. Therefore, it would be possible to tell if the pregnancy was at low risk or high risk.

The counselor spent some time discussing what Sarah and Nick would do if the pregnancy was at high risk. If the pregnancy continued and Sarah then developed Huntington's disease the baby would have been shown to have inherited the gene, because it would then be known that marker B was linked to the Huntington's disease gene in Sarah's father. This would be a very difficult situation for that child to be in and not at all desirable. It was also important for Sarah and Nick to understand that if the pregnancy was high risk and was terminated and Sarah was then shown not to have the gene for Huntington's disease, then they would have terminated a healthy baby.

Sarah and Nick still thought that they would have the pregnancy tested and would terminate a pregnancy at high risk. In order to see if the test was possible a DNA sample was needed from Nick. For the linkage to be informative for the pregnancy it has to be possible to distinguish between the markers that the fetus had inherited from Sarah and Nick. The linkage test would also have a small error rate because of recombination. It would theoretically be possible for the markers even though they are tightly linked to the Huntington's disease locus to cross-over during meiosis.

Because Sarah was 8 weeks' pregnant they were given another appointment in 1 week's time to discuss all the results and go over the details of exclusion testing again. The results of the DNA typing are shown in Figure 6 together with the possible outcomes for the pregnancy.

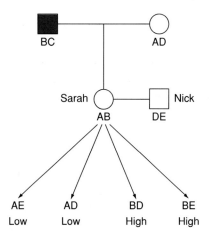

BC

AD

Sarah

AB

Nick

DE

AE
Low

AD
Low

BD
High

BE
High

Figure 6. Possible linkage results.

*Nick and Sarah came to the appointment, on their own this time. They were very
clear about their decision, which was to go ahead with testing. A CVS was booked
for 10 weeks of pregnancy. Sarah decided that she wanted to be phoned at home
with the result in the evening when they both would be home from work, and this
was agreed.*

*A week after the CVS was done the DNA result came through. It showed that the
pregnancy was at low risk. Sarah and Nick were delighted and continued with the
pregnancy. They wanted to be seen again after the baby was born to go over what
had happened and a follow-up appointment was arranged. As the conversation
with the genetic counselor was ending Sarah mentioned that she was very worried
about her sister Lauren. She had become anxious and depressed and seemed to
be a bit clumsy, but would not talk about it or see anyone. They agreed to talk
about this again at the follow-up appointment.*

KEY PRACTICE POINT

DNA analysis for prenatal diagnosis may require some work to be done in
the molecular genetics laboratory before it is possible to offer a test. It is
essential that the genetics laboratory, the clinical genetics team and the
obstetric and midwifery team liase closely when such testing is being
considered. Early referral (before pregnancy if possible) is also essential.

5.5 *Cordocentesis*

Occasionally, it may be desirable to sample some blood from the fetus. The sample is taken from the umbilical cord and the test can be done after 16 weeks' gestation. It can be used as a source of cells for karyotyping or for direct analysis of the blood, for example, for an assessment of anemia. It has a pregnancy loss risk of ~ 1.4% [34] (*Figure 7*).

6. Examination of the neonate

It is essential that any newborn is examined by the midwife or doctor, to detect any unusual or abnormal features at birth. Although this is important for a healthy infant, the results of the examination assume greater significance if the baby is stillborn or dies soon after birth. Couples who have lost a child around the time of birth are understandably shocked and distressed, and a future pregnancy may well be the furthest thought from their minds. However, when they begin to consider another pregnancy, one of their main concerns will be the possibility of recurrence.

It is impossible to offer accurate information about the recurrence risk without detailed records of the baby who did not survive. It is therefore vital that whatever the outlook of the parents at the time of birth, the midwife examines the baby thoroughly and carefully documents any unusual features. It is equally important to *document* normal findings, as this enables the genetic counselor to give a more informed opinion about the causes of death and chance of similar events occurring in the future.

A suggested checklist for the neonatal examination is included on the next page (*Figure 8*).

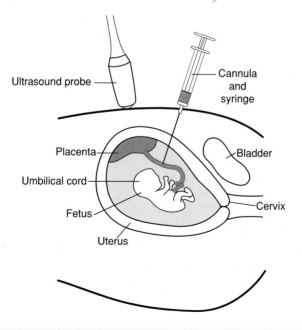

Figure 7. Cordocentesis.

Suggested checklist of features for examination of baby		
Gestational age		
Length		
Weight		
OFC		
Head	Shape	
	Fontanelles	
Hair	Texture	
	Color	
	Quantity	
Eyes	Brows	
	Lashes	
	Irides	
	Lens	
Mouth	Size	
	Shape	
	Lips	
	Palate	
Nose	Shape	
	Bridge	
	Nares	
Philtrum		
Mid-face		
Jaw		
Ears	Position	
	Size	
	Shape	
Chest		
Abdomen	Herniae	
	Genitalia	
	Testes	
Limbs		
Hands	Nails	
	Fingers	
	Creases	
	Size	
Feet	Nails	
	Toes	
	Hallux	
	Creases	
	Size	
Skin	Unusual patterns	
	Birthmarks	
	Skin tags	

Figure 8. Suggested checklist for the neonatal examination.

7. Post-mortem examination of a neonate

The subject of the post-mortem examination of the baby after a stillbirth or **neonatal death** is a sensitive one. Certainly pressure or coercion should never be applied to induce the parents to consent against their will. However, the decision should be an informed one, and therefore parents need information about the limitations on advice for the future if a post mortem is not conducted. The post-mortem examination provides accurate information about the cause of death and a detailed assessment of any external and internal abnormalities. Both are usually necessary for a genetic risk assessment to be made.

If the parents do not wish to consent to a post mortem, good quality photographs of the baby should be taken for the medical records, if possible by a medical photographer. These should include pictures of any abnormal features and the face, hands and feet. The parents may also consent to an external physical examination by a pathologist and/or X-rays. In addition, cord blood, venous blood or a skin sample can be sent for chromosome analysis. It is often possible to culture cells from the skin sample even if the baby is stillborn, although the culture is not likely to be successful if the fetus is significantly macerated. Amniocentesis performed before a termination of pregnancy may give the best opportunity for chromosome analysis on the fetus.

After spontaneous abortion or termination of pregnancy, products of conception can be sent in a container of normal saline or tissue culture medium for chromosome analysis.

KEY PRACTICE POINT

For obtaining a karyotype from a stillborn child, a small sample of skin can be taken from the upper leg using a skin biopsy punch or a small scalpel blade. The tissue should then be placed in tissue culture medium. Samples for karyotyping should NEVER be placed in formalin. The sample needs to be sent to the cytogenetics laboratory as a matter of urgency as delay lessens the chance that the culture will grow adequately.

Following a stillbirth, neonatal death or termination of pregnancy for fetal abnormality, parents may be reluctant to look at their baby, particularly if they are aware of congenital abnormalities. In fact, the parents may imagine features that are much more frightening than the actuality. In our experience the decision not to see the baby is often deeply regretted later. Skillful reassurance from the midwife may encourage the parents to see their baby, thus reducing later regrets and enabling the parents to grieve more appropriately. Wrapping the baby in a blanket will enable the parents to view their son or daughter in their own time if they wish to do so. Many parents are reassured by the normality of the majority of the features, it should be remembered that even babies with gross abnormalities such an anencephaly have a normal trunk, limbs, hands and feet.

If the parents do not wish to see their baby, consent should be sought to take some photographs. These can be kept securely in the notes and requested by the parents later. Handprints, footprints, or any item connected with the baby such as identity bands may be treasured by the parents. Such mementos help to reinforce the reality of the baby's existence, and may be helpful to the parents as they mourn.

8. Conclusion

In this chapter, we have provided some guidelines for the care of a mother who has a genetic condition, for a couple requesting genetic testing during pregnancy, and the examination of a neonate. The discussion questions at the end of this chapter are designed to help the reader think about the psychological issues in more depth. The next chapter of the book deals with genetic problems that may become obvious during infancy.

TEST YOURSELF – CASE DISCUSSIONS

Q1. Sally was a 22-year-old woman, pregnant for the first time. The ultrasound scan performed at 18 weeks showed the fetus had a heart defect, and an amniocentesis was done. Down syndrome was diagnosed in the fetus. Sally's partner was adamant he could not cope with an abnormal child and they decided to have the pregnancy terminated at 22 weeks. Sally was advised by her midwife not to see the baby as it would upset her.

Sally was tormented for several years by the thought that her daughter was too 'monstrous' to be seen. She also felt guilty that she had abandoned her. She had no photographs of the baby.

When she decided to try for another baby she was referred to a genetic counselor. Sally was very distressed when talking about her first baby. She expressed her regret at not seeing her daughter. Clinical slides had been taken prior to the post mortem. These were obtained by the genetic counselor and made into photographs. Sally was overwhelmed by the sight of her daughter, and was able to grieve for her, keeping the photos under her pillow. She has since had two more children.

Reflective practice: how can midwives or nurses present at the delivery of a baby with abnormalities discuss the baby's features realistically without creating fear in the parent? What terms could be used?

Q2. Karen was pregnant for the first time at the age of 38. She was fit and healthy and had no real worries about the pregnancy. She was offered amniocentesis and she and her partner decided to take up the offer to give them the reassurance that the baby was, as they put it, perfectly healthy.

A week after the amniocentesis was done, Karen's midwife called round to say that there was a problem with the result. The baby had a chromosome abnormality and although the midwife could not say any thing more about it she had made an appointment at the hospital with the consultant in 2 days' time. This was the earliest appointment available.

Karen and Michael spent the next 2 days trying to find out what a 'chromosome abnormality' was. They looked on the internet and found lots of information about trisomies, duplications and deletions. They saw pictures of children with severe learning disabilities and congenital abnormalities. They found out that if they wanted to have a termination of the pregnancy it would require an induction of labor and delivery of the dead baby. Karen and Michael realized that termination would be very difficult for them and in fact might be something they would not consider, but could they cope with a child with, as they now believed severe problems?

Their first words when they walked into the consultant's clinic were 'We only wanted a test to see if the baby was normal.' The obstetrician told them that the baby had a condition called Klinefelter's syndrome.

Males with Klinefelter syndrome have a 47XXY karytoype. They are generally tall, have small testes and are usually infertile. They may have enlarged breasts after adolescence (gynecomastia). Some boys with Klinefelter syndrome have learning problems; these are rarely severe [35].

Their reaction was initially one of relief but changed to a feeling of 'we still won't have our normal baby.' They believed intellectually that the baby was not going to have severe problems, but emotionally they felt that the baby was abnormal. The consultant emphasized that the baby would look normal and would be unlikely to have severe learning difficulties but did offer a termination of pregnancy. Karen and Michael decided to continue with the pregnancy. Karen became increasingly detached from the baby as the pregnancy advanced. When the baby was born he appeared perfectly healthy. The health visitor noted on her first visit that Karen seemed to have some difficulties in relating to the baby.

How could this situation have been handled differently?

Should the couple have been offered a termination?

What information could have been given to Karen and Michael before they decided to have an amniocentesis?

References

1. Colman JM, Sermer M, Seaward PG, Siu SC (2000) Congenital heart disease in pregnancy. *Cardiol Rev* **8** (3):166–173.
2. Brooten D, Kaye J, Poutasse SM *et al.* (1998) Frequency, timing, and diagnoses of antenatal hospitalizations in women with high-risk pregnancies. *J Perinatol* **18** (5): 372–376.
3. Lind J, Wallenburg HC (2001) The Marfan syndrome and pregnancy: a retrospective study in a Dutch population. *Eur J Obstet Gynecol Reprod Biol* **98** (1): 28–35.
4. Aburawi EH, O'Sullivan J, Hasan A (2001) Marfan's syndrome: a review. *Hosp Med* **62** (3): 153–157.
5. Lipscomb KJ, Smith JC, Clarke B, Donnai P, Harris R (1997) Outcome of pregnancy in women with Marfan's syndrome. *Br J Obstet Gynaecol* **104** (2): 201–206.
6. Bruno PA, Napolitano V, Votino F, Di Mauro P, Nappi C (1997) Pregnancy and delivery in Ehlers–Danlos syndrome type V. *Clin Exp Obstet Gynecol* **24** (3): 152–153.
7. Koch R, Hanley W, Levy H *et al.* (2000) Maternal phenylketonuria: an international study. *Mol Genet Metab* **71** (1–2): 233–239.
8. Sheard NF (2000) Importance of diet in maternal phenylketonuria. *Nutr Rev* **58** (8): 236–239.
9. Olson GL (1997) Cystic fibrosis in pregnancy. *Semin Perinatol* **21** (4): 307–312.
10. Tuck SM, Jensen CE, Wonke B, Yardumian A (1998) Pregnancy management and outcomes in women with thalassaemia major. *J Pediatr Endocrinol Metab* **11** (Suppl. 3): 923–928.
11. Aessopos A, Karabatsos F, Farmakis D *et al.* (1999) Pregnancy in patients with well-treated beta-thalassemia: outcome for mothers and newborn infants. *Am J Obstet Gynecol* **180** (2 Pt 1): 360–365.
12. Daskalakis GJ, Papageorgiou IS, Antsaklis AJ, Michalas SK (1998) Pregnancy and homozygous beta thalassaemia major. *Br J Obstet Gynaecol* **105** (9): 1028–1032.
13. Balgir RS, Dash BP, Das RK (1997) Fetal outcome and childhood mortality in offspring of mothers with sickle cell trait and disease. *Ind J Pediatr* **64** (1): 79–84.
14. Mahomed K (2000) Prophylactic versus selective blood transfusion for sickle cell anaemia during pregnancy. *Cochrane Database Syst Rev* **2**: CD000040.
15. Oakley CM (1997) Pregnancy and congenital heart disease. *Heart* **78** (1): 12–14.
16. Warburton D (1987) Chromosomal causes of fetal death. *Clin Obstet Gynecol* **30** (2): 268–277.
17. Boue J, Bou A, Lazar P (1975) Retrospective and prospective epidemiological studies of 1500 karyotyped spontaneous human abortions. *Teratology* **12** (1): 11–26.
18. Brambati B (1990) Fate of human pregnancies. In RG Edwards (ed.) *Establishing A Successful Human Pregnancy. Serono Symposia*, pp. 269–281. New York: Raven Press.
19. Harper PS (1998) *Practical Genetic Counselling*, 5th edn. Oxford: Butterworth-Heinemann.
20. Lee RM, Silver RM (2000) Recurrent pregnancy loss: summary and clinical recommendations. *Semin Reprod Med* **18** (4): 433–440.
21. Mueller R, Young ID, Emery A (1998) *Elements of Medical Genetics*, 10th edn. Edinburgh: Churchill Livingstone.
22. Williams C, Alderson P, Farsides B (2002) Is nondirectiveness possible within the context of antenatal screening and testing? *Soc Sci Med* **54** (3): 339–347.
23. Hook EB, Lindsjo A (1978) Down syndrome in live births by single year maternal age interval in a Swedish study: comparison with results from a New York State study. *Am J Hum Genet* **30** (1): 19–27.
24. Gilbert RE, Augood C, Gupta R *et al.* (2001) Screening for Down's syndrome:

effects, safety, and cost effectiveness of first and second trimester strategies. *BMJ* **323** (7310): 423–425.

25. Spencer K, Spencer CE, Power M, Moakes A, Nicolaides KH (2000) One stop clinic for assessment of risk for fetal anomalies: a report of the first year of prospective screening for chromosomal anomalies in the first trimester. *Br J Obstet Gynaecol* **107** (10): 1271–1275.

26. VanDorsten JP, Hulsey TC, Newman RB, Menard MK (1998) Fetal anomaly detection by second-trimester ultrasonography in a tertiary center. *Am J Obstet Gynecol* **178** (4): 742–749.

27. Luck CA (1992) Value of routine ultrasound scanning at 19 weeks: a four year study of 8849 deliveries. *BMJ* **304** (6840): 1474–1478.

28. Leschot NJ, Verjaal M, Treffers PE (1985) Risks of midtrimester amniocentesis; assessment in 3000 pregnancies. *Br J Obstet Gynaecol* **92** (8): 804–807.

29. Modell B (1985) Chorionic villus sampling. Evaluating safety and efficacy. *Lancet* **1** (8431): 737–740.

30. Firth HV, Boyd PA, Chamberlain P, MacKenzie IZ, Lindenbaum RH, Huson SM (1991) Limb abnormalities and chorion villus sampling. *Lancet* **338** (8758): 51.

31. Sundberg K, Bang J, Smidt-Jensen S *et al.* (1997) Randomised study of risk of fetal loss related to early amniocentesis versus chorionic villus sampling. *Lancet* **350** (9079): 697–703.

32. Alfirevic Z, Gosden CM, Neilson JP (2000) Chorion villus sampling versus amniocentesis for prenatal diagnosis. *Cochrane Database Syst Rev* **2**: CD000055.

33. Maat-Kievit A, Vegter-Van-Der-Vlis M, Zoeteweij M, Losekoot M, van Haeringen A, Roos RA (1999) Predictive testing of 25 percent at-risk individuals for Huntington disease (1987–1997). *Am J Med Genet* **88** (6): 662–668.

34. Tongsong T, Wanapirak C, Kunavikatikul C, Sirirchotiyakul S, Piyamongkol W, Chanprapaph P (2001) Fetal loss rate associated with cordocentesis at midgestation. *Am J Obstet Gynecol* **184** (4): 719–723.

35. Ratcliffe S (1999) Long-term outcome in children of sex chromosome abnormalities. *Arch Dis Child* **80** (2): 192–195.

Further reading

Abramsky L, Chapple J (editors) (1994) *Prenatal Diagnosis – The Human Side*. Cheltenham: Nelson-Thornes. (Thorough coverage of prenatal diagnosis and the effects on families.)

Kluger-Bell, K (1999) *Unspeakable Losses*. London: Penguin Books. (Discussion of the care of families who have lost a baby, giving insight into the emotional reactions of parents.)

McLachlan J (1994). *Medical Embryology*. Wokingham: Addison-Wesley. (Embryological text book with clear diagrams and explanation of abnormal development.)

7 Infancy

1. Introduction

It is important for the practitioner to remember that despite the majority of pregnancies ending with the birth of an apparently healthy child, in ~ 1–2% of pregnancies the baby is born with a significant congenital abnormality or disease [1]. In addition neonatal screening programs may detect a disease in an apparently healthy neonate.

The health professional at this time needs to be able to give and interpret accurate information, taking into account the distress and shock of the parents, and making an assessment of their particular needs. In medical terms the search may be for a diagnosis in order that treatment decisions may be made and information gathered about the prognosis. However, there may be considerable uncertainty about a diagnosis and it is possible that one will not be made at this time because important signs and symptoms may not yet be apparent. It is very understandable that parents will be searching for a diagnosis that explains the condition of their child, and the lack of certainty at this time can be very difficult. One important role of the health professional would be to help the family to focus on the needs of the child rather than the search for a diagnosis.

In this chapter we describe some of the more common congenital abnormalities, discuss neonatal screening programs and describe the possible diagnostic pathways for babies with chromosome abnormalities and neurological problems.

For the practitioner to work in partnership with the parents and the medical team, he or she needs to be informed and to understand the process. In this way they can truly care for the baby and their family as they come to terms with the difficulties they may be facing.

2. Congenital abnormalities

A congenital abnormality can be described as an alteration in the normal pattern of structural development, which is present at birth. Congenital abnormalities may be characterized according to the possible mechanism that caused them and whether they are isolated or form a pattern.

2.1 Cleft lip and palate

About 1 in 500 to 1 in 1000 children are born with a cleft lip or palate [2]. Although isolated cleft lip and/or palate are essentially treatable with surgery, there can be concern about the possibility of having a child with similar problems in the future.

In order to give accurate recurrence risks a careful examination must be made to exclude other anomalies that may indicate that the clefting is part of a syndrome and not isolated.

Examples of syndromes associated with cleft lip and/or palate:
- Van der Woude (autosomal dominant, lip pits with cleft lip/palate);
- Stickler syndrome (autosomal dominant, hereditary arthro-ophthalmopathy);
- trisomy 13;
- chromosome 22q deletion.

Teratogens such as anti-epileptic medication should also be considered as a contributory factor. This may change the management in a future pregnancy.

Studies have shown that cleft palate alone is a separate condition from cleft lip with or without cleft palate. Isolated cleft lip and/or palate are generally considered to be multifactorial/polygenic in etiology, although it is now thought that a proportion may be due to single genes. The empiric recurrence risks for future pregnancies after the birth of one affected child are between 2 and 4% [3–5].

CASE EXAMPLE   AMY

Amy was born at term and was noted to have a cleft palate. In addition to this she had a very small jaw. This caused her problems in maintaining her airway and she required initial ventilation. The diagnosis of Pierre Robin sequence was made. In this sequence the small jaw is thought to be the result of the failure of the tongue to influence the jaw development because it remains at the roof of the mouth due to the cleft palate. Amy's mother Debbie was put in touch with the Pierre Robin sequence support group and was able to talk to other families who had experienced the same as her.

The prognosis for the patient with Pierre Robin sequence is good with careful management and the recurrence risk is low [6]. Amy's mother was reassured by this, as her cousin, Faye was pregnant and due to have her baby at any time.

When Faye's baby was born she was noted to have a cleft uvula, bilateral dislocated hips and large round eyes. The geneticist who saw the family at that time found out that several members of the family were very short-sighted and that some of them had had detached retinas. In addition some members of the family had arthritis of the hips and knees. The geneticist made the diagnosis of Stickler syndrome. This is an autosomal dominant syndrome of cleft palate, high myopia with retinal detachment and arthropathy. The expression and penetrance is very variable. In addition the condition is **heterogeneous** with at least three genes having been identified, with many different mutations. All the genes identified code for collagen, a component of connective tissue [7].

After putting the family history together it became clear that Faye and Debbie both had the gene mutation for Stickler syndrome (*Figure 1*). Faye had always worn glasses and at school was a bit 'double jointed' but had had no other health problems. Debbie felt that Faye and herself were obviously from the same family, they took after their grandfather and looked more like sisters than cousins, but Faye had no medical problems. Their grandfather had had a hip replacement in his 50s and had in fact gone blind suddenly which family said was due to shock in the war.

Because Stickler syndrome is an autosomal dominant condition, both Faye and Debbie would have a 50% chance of passing the gene mutation to any of their children, however, how the gene would be expressed in any individual child could not be predicted. Debbie asked if the gene was 'stronger' in her as Amy had been quite severely affected. The different ways the gene affected people in the family is an example of variable penetrance and expression.

Apart from the way this diagnosis changed the information about prognosis and recurrence risks it also has important implications for management. The children and other members of the family now see an ophthalmologist regularly for review in order to monitor their eyes and to treat detached retinas in order to reduce the risk of blindness.

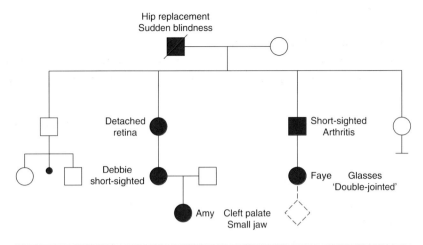

Figure 1. Family tree.

2.2 Neural tube defect

This term includes spina bifida, anencephaly and any other defect that is a failure of closure of the neural tube. The incidence of spina bifida varies between countries, regions of countries and also over time. A public health campaign demonstrated a reduction in the total number of births with a neural tube defect from ~ 2 in every 1000 children in 1996 to ~ 1.1 per 1000 children in 1999. Because spina bifida accounts for approximately half of all neural tube defects; this means the birth rate for spina bifida since 1996 has been between 0.5 and 0.65 per 1000 births [8].

When seeing a family for genetic counseling, as with cleft palate it is important to distinguish between cases in which the neural tube defect is isolated and those in which it forms part of a syndrome.

Recurrence risk for isolated neural tube defects is ~ 4%, but as discussed in *Chapter 4* this can be reduced by insuring an adequate intake of folic acid.

It should be noted that spina bifida and anencephaly are probably part of the same spectrum and the recurrence risk includes the whole spectrum. This may make a difference to how a family will perceive the recurrence risk.

Hydrocephalus is often associated with neural tube defects as a secondary consequence of disturbance of the circulation of the cerebrospinal fluid. There is a rare form of X-linked hydrocephalus associated with mutations in the *L1CAM* gene and hydrocephalus may also be part of other rare syndromes. Careful examination is needed to exclude neural tube defects before considering that a baby with hydrocephalus has one of these rare isolated forms.

2.3 *Gastroschisis*

This is a failure of closure of the abdominal wall, which allows the intestines to protrude through the wall. There is no evidence that this is a genetic condition and the recurrence risk is low. However, the incidence is increasing and the reasons for this are unclear.

There is some evidence that babies of younger mothers are at slightly greater risk [9,10].

2.4 *Congenital heart defects*

The recurrence risk for isolated congenital heart defects (CHD) varies according to the nature of the defect. With improvements in surgery children are surviving to reproductive age who would not have survived in the past. There is therefore little empiric information as to the risk of their children having a similar congenital heart defect and they may request genetic counseling prior to conception.

Many chromosome abnormalities are associated with congenital heart disease and chromosome analysis should be considered if there are any other associated abnormalities. The identification of a microdeletion on chromosome 22 that is associated with conotruncal cardiac abnormalities has clarified the etiology in a subset of patients and detailed chromosome analysis should be considered if appropriate.

These microdeletions are found in more than half of children born with interrupted aortic arch or truncus arteriosus. Identification of such a deletion has implications for recurrence risk and also for the presence of associated abnormalities [11].

CASE EXAMPLE   **KIMBERLEY**

Kimberley had been diagnosed with an interrupted aortic arch shortly after birth. Her parents were with her on the pediatric cardiology ward waiting for her assessment and decisions to be made about potential surgery. As part of the routine workup Kimberley's chromosomes had been checked and a small deletion on chromosome 22q had been found. This was discussed with her parents and Kimberley was carefully examined to make sure her palate was intact, she also had her calcium levels checked as babies with a 22q deletion may have subtle palatal abnormalities or abnormalities of calcium metabolism. When the phenotype of 22q deletion was discussed with Kimberley's parents her mum mentioned that she (mother) really had not settled in at school and that the other children made fun of her voice saying she spoke through her nose. She also felt that she was always getting colds and infections and that although she had 'done okay' at school, she had not done quite as well as the rest of the family. When her chromosomes were checked she was also shown to have the deletion but the effect on her was very different to that on Kimberley.

---

3. Chromosome abnormalities

If a baby is born with a combination of congenital abnormalities, chromosome analysis is normally considered. As discussed in *Chapter 4*, karyotyping can be carried out at varying levels of detail. For example, a rapid karyotype reported in 24 hours may be sufficient to exclude trisomies such as Down syndrome or trisomy 18, but may not detect subtle deletions or rearrangements.

Studies of unselected newborn populations [12] show that ~ 1% of new-borns had recognizable chromosome abnormalities, about three-quarters of which were autosomal trisomies and the remainder involved the sex chromosomes.

The most common chromosome abnormality that the health professional will encounter in practice is Down syndrome – trisomy 21, although there are many other possible chromosome abnormalities that can be identified.

Inskipp family

The following referral was made by phone to the genetics department from the special care baby unit.

Maria Inskipp
Born today Cesarean section
Small for dates
Cleft lip and palate
Microcephaly
VSD
Please see and advise

The clinical geneticist visited the neonatal unit. The baby's father, David was there; her mother, Sally had not recovered sufficiently from the emergency Cesarean section to visit her baby yet.

The findings that were noted on examination were:
Bilateral cleft lip and palate
Microcephaly
Hypotonia
Highly arched eyebrows
Hypertelorism

The baby had already had a blood sample taken for karyotyping and an initial result was expected in 48 hours. An arrangement was made to see Sally and David together within the next few days.

The clinical findings in Maria suggested that Maria might have a condition called Wolf–Hirschorn syndrome. This is known to be associated with a deletion of the short arm of chromosome 4 (4p–) (see Chapter 4, Section 3 for an explanation of karyotyping). However, the initial result of Maria's chromosome analysis was reported as normal. The geneticist also took a family history from David. Maria was their first baby and they had not had any other pregnancies.

Many hundreds of congenital abnormalities are described involving all body systems. For further detailed information see books by Gorlin or Jones, or the OMIM website. The role of the genetic team is to try to determine whether the abnormality is isolated or part of a wider 'syndrome' or 'association'. This at the very least requires a detailed examination, and may require further diagnostic tests such as karyotyping.

4. Examination of the neonate

Diagnostic evaluation of a baby born with congenital abnormalities is the responsibility of a medical practitioner, but the nurse or midwife caring for the family needs to have an understanding of the components of the process in order to work with the family and to be able to initiate appropriate referral if necessary.

Components of diagnostic evaluation include [13]:
History:
- prenatal;
- perinatal;
- family history.

Physical examination:
- assessment of growth;
- general appearance;
- detailed examination.

The history and examination will lead to an initial impression and differential diagnosis. This will guide subsequent diagnostic tests. A detailed check list for examination of a baby can be found in *Chapter 6*.

Inskipp Family

The geneticist thought a diagnosis of Wolf–Hirschorn syndrome was possible in Maria. This is associated with a deletion of the end of the short arm of chromosome 4. She therefore asked the cytogenetic laboratory to use a FISH probe to establish the presence or absence of a deletion. (See Chapter 4 for an explanation of FISH.)

The cytogenetic analysis confirmed the diagnosis in Maria and also showed that Maria had an unbalanced translocation. David and Sally were seen together and told about the diagnosis. Their first question was 'How long will she live?'. Older reports of children with this syndrome implied that many of them would die within the first 2 years of life. However, recent reports suggested that average survival might now be longer [14]. This is probably because of improvements in medical treatments, but will also be the result of more diagnoses being made because of improvements in cytogenetic techniques.

5. Breaking bad news

It is important to realize that the way the family is told any possible diagnoses and how they are subsequently managed is only the first step in an ongoing interaction with the medical profession. As with the breaking of any difficult news there are a number of issues to consider.

5.1 The environment

Ideally, a quiet private environment should be provided. If desired by the family, other family members should be included.

5.2 The clinical encounter

The attitude of the health professional should be supportive, nonjudgmental and take account of the cultural and ethnic values of the family.

There should be respect for the autonomy of the family and, of course, for their confidentiality.

It should be recognized that the family will go through a form of grieving and this needs to be taken into account.

5.3 Content

The medical facts as they are known should be given, which should include possible causes, the diagnosis and prognosis.

Realistic prognoses if possible should be given but include optimism if appropriate.

The content should be tailored to the family's need to know and it should be emphasized that there will be further opportunities for discussion and clarification.

Extensive opportunities should be given for the family to ask questions and express their concerns.

At all times the clinician should be honest and truthful about what they know and do not know.

Information on patient support groups and other support services should be made accessible.

It should be recognized that immediately after the death of a child or the diagnosis of a serious problem that the parents might be unable to receive or understand complex information. Knowing how and when to give information requires considerable skill on the part of the health professional.

The birth of any baby with an abnormality within a family will affect all the members of that family. Grandparents in particular may find it difficult to know how best to help both their child and grandchild and while being a

KEY PRACTICE POINT

Information about rare conditions may be based on studies that were done some time ago and prognoses may have changed because of advances in medicine and diagnosis. It is essential to get as much up to date information as possible before giving advice on the prognosis of any condition.

Parents may search for information in libraries or on the Internet. Cases that are included in publications tend to be those that are at the more severe or unusual end of the spectrum, and parents should be warned about this and the fact that mild cases are frequently not diagnosed and therefore not reported.

Inskipp Family

Because Maria had been shown to have an unbalanced rearrangement Sally and David also had their chromosomes checked. Sally was shown to have a balanced translocation involving chromosome numbers 4 and 6. This balanced translocation would have implications for the rest of the family and it was recommended that Sally tell her family and arrange for her parents to have their chromosomes checked. Sally said she could not do this, her parents, particularly her mother, had been so distressed by what had happened that Sally felt they could not take on any more. They could not understand what had happened and wanted to blame somebody, they also were full of guilt and worry about Sally. Sally felt that if her mother were shown to have the translocation she would not be able to come to terms with it and would blame herself. Sally and David wanted to concentrate on caring for Maria and looking after each other.

source of support for the parents of the baby, may also feel excluded from the support and advice that is available for the parents. This is starting to be recognized by patient support groups who may offer specific advice and support to people in this situation [15]. This can be brought sharply into focus when a diagnosis in a child has implications for the rest of the family. The health professional needs to be sensitive to these issues when caring for the family.

6. Genetic testing in the neonate

In the same way that increasingly detailed karyotyping in the neonate allows for more diagnoses to be made at an earlier stage, the addition of molecular genetic techniques to careful clinical and pathological examination has increased the understanding of a diverse, but individually rare, group of neural, muscle and metabolic disorders that may present in the neonatal period. As before, an accurate diagnosis will allow specific information about prognosis and implications for other family members to be given.

6.1 *Examples of conditions presenting in the neonatal period with muscle weakness*

Congenital myotonic dystrophy	autosomal dominant maternally transmitted
Type 1 spinal muscular atrophy (Werdnig Hoffman)	autosomal recessive
Centronuclear (myotubular)	X-linked myopathy
Mitochondrial myopathies	heterogeneous dominant recessive or mitochondrial
Prader–Willi syndrome	chromosomal

The addition of molecular genetic techniques to the pathological and clinical examination may lead to diagnoses being made earlier than otherwise would have happened in the past. As discussed earlier in this chapter, the nurse who is caring for the family needs to insure that any information she gives to the family is accurate and represents the current best knowledge.

CASE EXAMPLE   BABY LANG

Baby Lang was born at term, he was a normal birth weight but had very poor muscle tone and sucking reflex. He was not dysmorphic and had no relevant family history. The pediatrician examined him and considered the possible diagnoses. He considered asking for a neurological opinion and also asked for chromosomes to be checked. When the laboratory received the specimen they noted that hypotonia was mentioned on the referral card and ensured that the karyotype included a detailed examination of chromosome 15. A small deletion was detected at 15q11 in the Prader–Willi critical region. Parents' chromosomes were requested and were reported as normal. However, DNA probes were used to identify whether the deletion had arisen on the maternal or paternal chromosome. The analysis confirmed that the deletion was on the chromosome that had been inherited from the father. Prader–Willi syndrome is caused by the loss of the paternal contribution in a critical region of chromosome 15 . This can happen through a variety of mechanisms, a deletion arising on the father's chromosome, inheritance of two copies of the father's chromosome 15 and none of the mother's, or by inheriting a gene that switches off that specific area on the father's chromosome 15. (For a more detailed discussion of imprinting see *Chapter 4*.)

In Prader–Willi syndrome the history that is normally given is of a baby who is hypotonic at birth and then has feeding difficulties. However, at about the age of 2 the child will start gaining weight and will show an extreme interest in food. The weight gain is not only due to overeating, people with Prader–Willi syndrome seem to be lacking in the normal appetite control mechanisms and also have a lower metabolic rate with a high fat to muscle ratio. Children with the syndrome have mild to moderate learning difficulties and although many of them start their education in mainstream school, they will normally need some educational help eventually [16,17].

As the genetic basis of diseases are clarified and more specific diagnostic and prognostic tests become available, the health professional will need accurate information and guidance to utilize this technology for the improvement of health care.

7. Neonatal screening

Screening is a process in which all individuals in a population which does not necessarily believe itself to be at risk of (or affected by) a condition are offered a test or asked a question to identify that condition. The aim of any screening program is to do more good than harm and reduce the risks of morbidity associated with specific conditions. Neonatal screening started in most Western countries in the late 1960s with screening programs for phenylketonuria (PKU). Screening for congenital hypothyroidism was then added [18]. Recently, in the UK universal neonatal screening for cystic fibrosis and the hemoglobinopathies has been adopted as policy.

7.1 Phenylketonuria

PKU is an autosomal recessively inherited inborn error of metabolism in which affected individuals are unable to metabolize the amino acid phenylalanine to tyrosine. This leads to high levels of phenylalanine that are neurotoxic. In the absence of treatment nearly all affected individuals develop severe, irreversible learning difficulties together with neurological deterioration. These severe manifestations are now very rarely seen because of universal neonatal screening and subsequent dietary restriction. Classical PKU is caused by a deficiency of the enzyme phenylalanine hydroxylase (PAH), varying degrees of PAH deficiency, as well as disorders of other enzymes in the metabolic pathway, all can lead to high levels of phenylalanine (*Figure 2*). The disorder is therefore heterogeneous.

All babies should have their phenylalanine levels measured on a dried blood spot (the **Guthrie test**). This should be done between 6 and 14 days after birth. Early diagnosis allows the initiation of a diet low in phenylalanine. In effect, this means that protein intake is extremely restricted and supplementary amino acids are taken. Recent data suggest that this diet should continue at least until adulthood. There are special concerns about

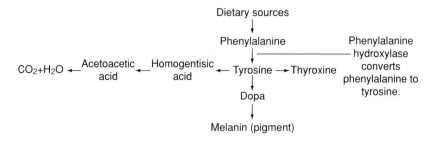

Figure 2. Enzyme pathway.

pregnancy in women affected with PKU and it is essential that low serum phenylalanine levels are maintained prior to conception and throughout the pregnancy. (See *Chapter 6* for a discussion of the management of the pregnant woman with PKU.) Although PKU is relatively rare (1 in 12 000 births) the fact that the prognosis, if untreated is so devastating, that there is an effective treatment and a screening test with high sensitivity and specificity make compelling arguments for universal **population screening** [19].

CASE EXAMPLE   **KATY**

Katy was just about to leave the postnatal ward after having her second baby when the midwife asked her to wait a few moments while they did the Guthrie test. Katy knew that all babies had a heel prick test just to check they were normal and was happy for this to be done. A week later she was contacted by a specialist nurse from the PKU service and told there was something wrong with the heel prick test and her baby needed to be seen and the test repeated on a blood sample. Katy was given some information about PKU but was also told that it was possible that this first abnormal test result may mean nothing. She was given an appointment to see the pediatrician. The result of the second test confirmed the raised phenylalanine levels and PKU was diagnosed. When the pediatrician discussed PKU with Katy and told her her baby was affected and would need to be on a special diet until he grew up, Katy was astonished. She had been given a leaflet about the heel prick test but she had no idea of the implications of that routine test that all babies have. The pediatrician also told her that the condition was genetic and that there was a one in four chance that any future children she had with the same father would also have the condition. She could not understand this as nobody in her family had this condition. The pediatrician explained that she and the baby's father carried one copy of this faulty gene but that by chance her son had inherited two copies and therefore could not make the enzyme necessary to convert phenylalanine to tyrosine, allowing the levels of phenylalanine to build up and affect the way her son's brain developed.

She was made an appointment to see the rest of the team that cared for children with PKU and the dietician involved in the team went through the sort of diet her baby would need including special formula milk. He would also need regular blood spot tests to check his levels of phenylalanine and to monitor his diet. Although Katy was reassured that treatment for her baby was possible, she was also worried about how she would manage preparing different foods for her two children and what she would do in the future about explaining the condition to her son. Would he cope as a teenager? Would she be able to afford the special food he might need? Would she be able to do his blood spot tests?

KEY PRACTICE POINT

Although information is given before population screening tests, a positive test result will still be unexpected. When giving positive screening tests results the practitioner should insure they have enough information and are able to refer to the appropriate specialist service as soon as possible.

7.2 Congenital hypothyroidism

Screening for congenital hypothyroidism is also performed on all neonates on the Guthrie blood spot. About 1 in 4000 babies are born with congenital hypothyroidism, 90% of whom will need lifelong treatment with thyroid hormone replacement. A functioning thyroid is required for normal growth and brain development. As with PKU, if congenital hypothyroidism is diagnosed early enough treatment is effective. Unlike PKU, the condition can be caused by:

- failure of the thyroid gland to form normally. This is normally sporadic not genetic (80–85% of cases);
- an abnormality of the structure or function of the hypothalamus and/or pituitary. There are a variety of rare medical conditions associated with this central hypothyroidism, some of which are structural, some genetic (< 5% of cases);
- an abnormality of the enzyme pathway necessary for thyroid hormone production or release. As with many inborn errors of metabolism the majority of these are autosomal recessive genetic disorders (10–15% of cases).

7.3 Hemoglobinopathy and cystic fibrosis screening

In 2001, the UK government announced that universal neonatal screening for cystic fibrosis would be introduced together with **hemoglobinopathy** screening. These programs are being developed and will use the Guthrie blood spot samples. There are a number of issues, which need to be taken into account in the design of these programs. For example, if DNA-based technology is used, carriers of these recessive genes will be detected as well as affected individuals. Should the parents of the baby be informed of their carrier status? The hemoglobinopathies are more prevalent in specific ethnic minorities, should screening be targeted to those ethnic minorities or be universal? The same issue applies to cystic fibrosis screening which is more prevalent among northern European populations. There are also issues about the design and delivery of these screening programs.

At the beginning of the 21st century the impact of genetics is moving out from the specialist services into general health care. In order to evaluate the benefits or otherwise of possible healthcare interventions the healthcare professional needs to be informed. This is important in relationship to screening programs, which involve taking a population who does not believe itself to be at risk and perhaps diagnosing a serious condition. In addition, all screening programs will make both false-positive and false-negative diagnoses, i.e. show a high risk of a specific condition that is not there or show a low risk of a condition that is in fact present. These problems are the same for all screening programs, whatever technology they are based on.

7.4 Other neonatal screening programs

The two national neonatal screening programs that have been in place the longest in the UK are for PKU and congenital hypothyroidism. Other screening programs for galactosemia, medium-chain Acyl-CoA dehydrogenase deficiency MCADD deficiency, and Duchenne muscular dystrophy are performed in other countries or as pilots in specified regions within the UK.

8. Conclusion

The impact of the diagnostic and prognostic information that will become possible through advances in genetics also requires the health professional to continue to gain the knowledge to apply these advances for the benefit of their patient. The issue of early diagnosis and the consequent uncertainty about prognosis has been discussed repeatedly in this chapter in relation to infancy. However, the increasing power of genetic advances to diagnose and predict is relevant at all stages of life as demonstrated in the chapters that follow.

TEST YOURSELF

Q1. What is the difference between genetic testing and population genetic screening?

Q2. A neonate is diagnosed as having a sub-microscopic-deletion of chromosome 22q11.

(a) Name the laboratory tests that may have been done to identify the micro-deletion.

(b) The parents plan to have another child and ask you about the risks of recurrence, how will you go about providing them with accurate information?

Q3. What is the difference between a Malformation and a Deformation?

References

1. Royal College of Physicians (1989). *A Report of the Royal College of Physicians: Prenatal Diagnosis and Genetic Screening.* London: Royal College of Physicians.
2. Harper PS (1998) *Practical Genetic Counselling*, 5th edn. Oxford: Butterworth-Heinemann.
3. Tenconi R, Clementi M, Turolla L (1988) Theoretical recurrence risks for cleft lip derived from a population of consecutive newborns. *J Med Genet* **25** (4): 243–246.
4. Carter CO, Evans K, Coffey R, Roberts JA, Buck A, Roberts MF (1982) A three generation family study of cleft lip with or without cleft palate. *J Med Genet* **19** (4): 246–261.
5. Carter CO, Evans K, Coffey R, Roberts JA, Buck A, Roberts MF (1982) A family study of isolated cleft palate. *J Med Genet* **19** (5): 329–331.
6. van den Elzen AP, Semmekrot BA, Bongers EM, Huygen PL, Marres HA (2001) Diagnosis and treatment of the Pierre Robin sequence: results of a retrospective clinical study and review of the literature. *Eur J Pediatr* **160** (1): 47–53.
7. Snead MP, Yates JR (1999) Clinical and molecular genetics of Stickler syndrome. *J Med Genet* **36** (5): 353–359.
8. Chan A, Pickering J, Haan E *et al.* (2001) 'Folate before pregnancy': the impact on women and health professionals of a population-based health promotion campaign in South Australia. *Med J Aust* **174** (12): 631–636.
9. Rankin J, Dillon E, Wright C (1999) Congenital anterior abdominal wall defects in the north of England, 1986–1996: occurrence and outcome. *Prenat Diagn* **19** (7): 662–668.
10. Byron-Scott R, Haan E, Chan A, Bower C, Scott H, Clark K (1998) A population-based study of abdominal wall defects in South Australia and Western Australia. *Paediatr Perinat Epidemiol* **12** (2): 136–151.
11. Ryan AK, Goodship JA, Wilson DI *et al.* (1997) Spectrum of clinical features associated with interstitial chromosome 22q11 deletions: a European collaborative study. *J Med Genet* **34** (10): 798–804.
12. Jacobs PA, Lelville M, Ratcliffe S, Keay AJ, Syme JA (1974) A cytogenetic study of 11,680 newborn infants. *Ann Hum Genet* **37**: 359–376.

13. American College of Medical Genetics (1999) Guideline: evaluation of the newborn with congenital anomalies. http://www.faseb.org/genetics/acmg
14. Shannon NL, Maltby EL, Rigby AS, Quarrell OW (2001) An epidemiological study of Wolf–Hirschhorn syndrome: life expectancy and cause of mortality. *J Med Genet* **38** (10): 674–679.
15. Contact a Family (2002) CaF factsheet: grandparents. http://www.cafamily.org.uk/grandparents.html
16. Holm VA, Cassidy SB, Butler MG *et al.* (1993) Prader–Willi syndrome: consensus diagnostic criteria. *Pediatrics* **91** (2): 398–402.
17. Greenswag LR (1987) Adults with Prader–Willi syndrome: a survey of 232 cases. *Dev Med Child Neurol* **29** (2): 145–152.
18. European Society of Human Genetics: Public and Professional Policy Committee (2002) Population genetic screening programmes: principles, techniques, practice and policies. http://www.eshg.org/screening.htm
19. NIH (2002) Phenylketonuria (PKU): screening and management. *NIH Consensus Statement* **17** (3): 1–33.

Further reading

Gorlin RJ *et al.* (2001) *Syndromes of the Head and Neck.* Oxford: Oxford University Press. (Detailed text on dysmorphic syndromes.)

Harper PS, Clarke AJ (1997) *Genetics Society and Clinical Practice.* Oxford: BIOS. (Discussion on screening.)

Jones, KL (1997). *Smith's Recognizable Patterns of Human Malformation*, 5th edn. Philadelphia: W.B. Saunders. (General text on dysmorphic syndromes with excellent description of formation of dysmorphic features.)

8 Childhood and adolescence

1. Introduction

This chapter is mainly concerned with those genetic conditions that become evident during childhood. In many cases, a genetic condition is first suspected when a child fails to reach their developmental milestones. As the health visitor often has the closest contact with the family, and is responsible for monitoring the child's development, a number of referrals come from that source.

However, if a child has a genetic syndrome, learning delay is rarely the only sign of the condition. A significant change in either the chromosome structure or single gene will almost invariably also have some physical manifestations. These may be striking, such as a cleft lip, or very subtle, for example, a double crown or small fingernails.

Physical characteristics that differ from the norm are termed 'dysmorphic features'. However, very few of us are completely perfectly made, and if you examine most children or adults you will find one or two mild dysmorphic features. In most families, there are significant physical characteristics that are the norm for those families, so it is always important to view a child in the context of their particular family.

> **KEY PRACTICE POINT**
>
> When talking to a family about the physical characteristics of a child, the family's view of the features is important. What may be unusual to the professional, may be 'just like Mum or Dad' to the family.

2. Why seek a diagnosis?

In some ways, finding the diagnosis may change very little when a child has health, educational or social problems due to a genetic condition. The genes are not changeable, and treatment is usually on a symptomatic basis. It could be argued that a child who has needs for therapy ought to have the therapy regardless of diagnosis. However, there are four key reasons for pursuing a genetic diagnosis.

▪ The parent's need for information about the child. It has been demonstrated that parents who have a child with health or educational problems search for the reasons for those problems [1]. Without a diagnosis it is often difficult or impossible to have sufficient information about the likely prognosis, and questions about the child's future health or development cannot be answered. Some parents may even withdraw from a child emotionally, to protect themselves from further distress if they are unsure about the child's long-term survival.

▪ Screening for complications. If a child has a genetic condition, they may be susceptible to complications or future health problems that could be avoided or treated promptly if the possibility of them occurring was known. For example, a child who had neurofibromatosis should have regular medical checks because they are susceptible to scoliosis and malignancy.

▪ Facilitating access to support. Although support for a child should be provided where there is need, in reality the diagnostic label often facilitates the parents in obtaining additional educational, social or financial support for their child.

▪ Genetic risk assessment for other family members. Without a definite diagnosis in the affected child, it is difficult to assess the level of risk to other family members, including other current or future children of the parents. The parents may wish to have prenatal diagnostic testing in a future pregnancy, and this would not be possible without diagnosis.

From a different perspective, having a diagnosis can lead to 'labeling' of the child. When a diagnostic label is given, the child may be judged according to reports of other children with the same condition, rather than on their own abilities. This may limit them in reaching their own potential, or more may be expected of them than they can achieve. If the term syndrome is used in the diagnostic context, this may lead to misunderstanding outside the medical profession, as many people think only of Down syndrome when they hear the word syndrome.

CASE EXAMPLE NATHAN

Nathan was the second child of Bob and Sue. Their first child, Helen, was a lively 7-year-old when Nathan arrived, and doing very well at school. Nathan was a difficult child from the start, he wouldn't feed properly, and never slept for more than 2 hours. He seemed 'floppy' to his mother, in comparison to Helen. Nathan was very slow to smile, and didn't even try to roll over on his own until he was 8 months old. Sue and Bob were concerned, but other people reassured them that 'boys are always slower than girls', and as he was quite a large baby he was bound to be 'lazy'. At 12 months, he was not able to sit alone, and he was referred to the pediatrician by his health visitor. The pediatrician felt he was delayed, and did a series of tests, including a metabolic screen and chromosome analysis. No abnormalities were found on either.

By the time Nathan was 18 months, it was clear his development was severely delayed. He was still not able to sit properly without cushions. He was referred to the genetic service for further investigations, but as he had few dysmorphic features, finding a diagnosis was not possible at the time. Bob and Sue spoke to their health visitor about the agony of waiting for information about the cause of Nathan's problems. The uncertainty was making it harder for them to accept his condition, and also made it impossible for them to have information about his future development and prognosis. Sue said 'It's like waiting, waiting, waiting every day, but I don't know what I'm waiting for, I'm just waiting. If I knew what we had to deal with, that would be easier, even if it was bad news, much easier to deal with that than the waiting.'

3. Developmental delay

Developmental delay is defined as a delay in reaching the normal milestones within the normal age range for each milestone, for example, failure to walk before the age of 18 months. Development tasks are usually divided into categories, related to motor tasks, speech and cognition, and children may be delayed in their development globally (in all three areas) or specifically. For example, it is not uncommon to see children who are slower than normal in attaining speech but who are progressing normally in other respects.

If delay is suspected, the child's development may be assessed formally using a recognized developmental test. This type of tool requires the child to complete a series of tasks; each designed to assess a different aspect of development, in the motor, speech or cognitive sphere. The tasks are assigned different points, and the number of points compared with that expected for the child's age.

EXAMPLE OF DEVELOPMENTAL ASSESSMENT TOOL

To assess manipulative skills, the child is asked to build a tower of bricks, with points ranging from 1 to 4 for the number of bricks used.

In assessing vocalization, the child who used one word meaningfully would be allocated one point, whereas a child using several words with meaning would be allocated 4 points.

Locomotor skills involving the use of stairs would be judged, with a child who crawls upstairs attaining 1 point, and one who is able to run upstairs being allocated 6 points.

Parents will frequently have been the first to be concerned about the child, and may have raised their concerns previously. However, sometimes a parent will not have detected the delay, because of inexperience or avoidance, and so the news that the child is delayed may come as a shock. In a family where learning problems have been experienced by a number of family members, the child's situation may be considered to be the norm.

4. Learning disability

Children who have a learning disability are usually able to learn, but do so at a slower rate than peers of the same age. This is important to explain to parents, who may feel their child will never develop the skills necessary for daily living.

Learning disabilities can be further defined.

Communication:
- expressive language disorder – difficulty using correct language;
- articulation disorder – difficulty making speech sounds;
- receptive language disorder – difficulty interpreting language.

Academic skills:
- dyslexia – difficulties reading the written word;
- writing disorder – difficulties expressing through the written word;
- arithmetic disorder – difficulties with number processes.

Motor skill disorders:

Attention disorders:
- inability to focus attention is often accompanied by hyperactivity; the combination is termed attention deficit hyperactivity disorder (ADHD).

The reasons for learning disability are varied, and frequently there is no detectable genetic reason. However, the number of cases that appear to 'cluster' in families, and the genetic influences on brain development lead inevitably to the conclusion that there is a genetic influence in many cases of learning disability. However, it is often difficult to make a firm genetic diagnosis unless the child also has significant dysmorphic features.

Each child needs to be assessed formally and provided with educational support relevant to his/her own individual needs. Frequently the involvement of a community pediatrician or educational psychologist will be helpful.

5. What is dysmorphism?

Dysmorphism is defined as an unusual pattern of physical features. Although there is a huge range of 'normality' in human characteristics, an abnormal gene or chromosome structure often alters the physical features in that child, beyond the limits of the normal range [2]. Although these unusual characteristics are themselves usually completely benign, they provide clues as to the gene or chromosome abnormality in that child. Children who share the same genetic abnormality will share common characteristics, and although unrelated, may look very similar to one another. The most well known example of children who have the same chromosome abnormality looking similar is probably Down syndrome. Most people would be able to identify a person with Down syndrome, because of the similar facial and body characteristics of those who have inherited an additional chromosome 21. Children with Down syndrome will also inherit some particular physical features that identify them as belonging to their family.

5.1 *Common dysmorphic features*

When a child with learning delay is referred to the genetic clinic for diagnosis, a thorough physical examination is essential, to detect any dysmorphic features that may give clues as the diagnosis [2]. A systematic examination (usually starting with the head and working down) is done, so small clues are not missed. Although there are hundreds of different dysmorphic features, some of the most common are:

Head
- Microcephaly –head size below 3rd centile.
- Macrocephaly – head size above 97th centile.
- Hydrocephaly – head size increased, due to excess fluid in ventricles.
- Delayed closure of fontanelles.
- Flat or prominent occiput.
- Craniosynostosis – abnormal joining of the bones of the skull, resulting in abnormal head shape.

Hair
- Abnormally thick or thin hair.
- Double crown.
- Widow's peak.
- Sparse hair.

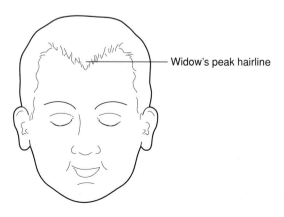

Widow's peak hairline

Figure 1. Widow's peak.

Eyes

- Hypotelorism – short space between eyes.
- Hypertelorism – long space between eyes.
- Slanting palpebral fissures (eye opening).
- Epicanthic folds – folds of skin at inner canthus of the eye.
- Prominent eyes.
- Microphthalmia (small eye) or anophthalmia (absence of eye).
- Blue sclera.
- Coloboma – 'gap' in iris.
- Cataract.

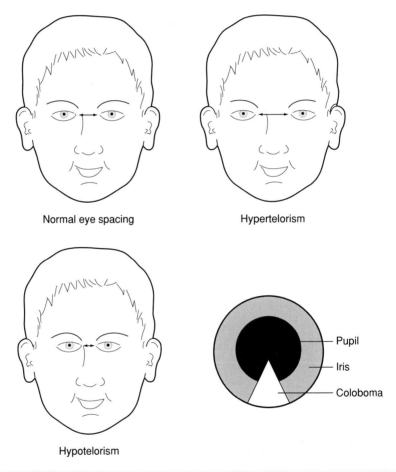

Normal eye spacing   Hypertelorism

Hypotelorism

Figure 2. Eye features.

Mouth

- Clefting of lips or palate.
- Prominent lips.
- Lip pits.
- Macroglossia.
- Hypoplasia of teeth enamel.
- Small or abnormally shapen teeth.
- Irregular placement of teeth.

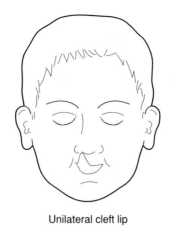

Unilateral cleft lip

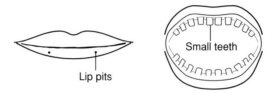

Lip pits

Small teeth

Figure 3. Mouth features.

Ears
- Malformation of auricles.
- Low-set ears.
- Pre-auricular tags or pits.

Hands/feet
- Brachydactyly.
- Clinodactyly.
- Hypoplasia of thumb or fingers.
- Hypoplasia of metacarpals.
- Polydactyly.
- Syndactyly.
- Broad thumb or toe.

Genitalia
- Hypospadius.
- Undescended testes.
- Unusually large or small penis/testes.
- Hypoplasia of labia majora.
- Vaginal atresia.

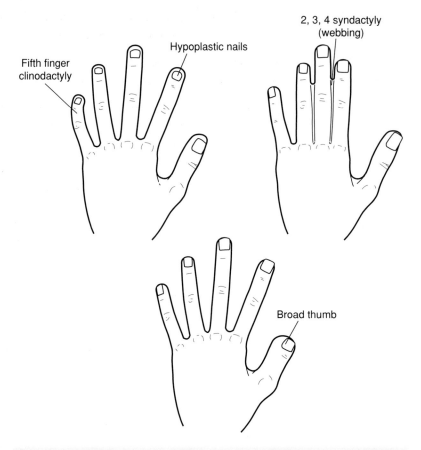

Figure 4. Hand features.

Stature
- Stature above 97th or below 3rd centile for age.
- Disproportion between trunk and limbs.

Skin
- Excessive or inadequate sweating.
- Altered skin pigmentation.
- Hemangiomata.
- Thick or ichthyotic skin.

Spine
- Neural tube defect.
- Scoliosis.
- Kyphosis.

CNS
- Hypertonicity.
- Hypotonicity.
- Seizures.

Additional information may relate to 'internal' abnormalities, such as a heart defect, renal or ureteric anomalies, or tracheo-esophageal fistulae.

Further information on particular dysmorphic features can be found in any recognized textbook on dysmorphology or genetic syndromes.

It is very common for photographs to be taken of children in the clinic, to act as an aid to memory at a later stage. These form part of the child's medical records, and consent is always sought from the parents (and child if old enough) before the photographs are taken.

Before and after a consultation, information about the child will be used to try and match the pattern of features in the child with those observed in other children. Increasingly, this is done using computer software packages, such as the London Dysmorphology Database (see website) or POSSUM. This type of software allows the principle features of the child to be listed, and a list of potential diagnoses that would be consistent with that pattern of features is produced. Of course, there is huge overlap in the patterns, so the skills of the geneticist are needed to differentiate the very probable from the less probable diagnoses. Inexperienced practitioners can look at a list of features and conclude that the child has a number of them, therefore a certain diagnosis must be correct. However, it is necessary to view the features as part of an overall appearance (or 'gestalt'), rather than as a number of unconnected features. The ability to assess a child in this way is attained with considerable clinical experience.

If the doctor seeing the child is unsure of the diagnosis, the case may be discussed in a wider forum, to access a greater range of clinical expertise. This may take the form of a team meeting, a regional meeting or even a national meeting for the purpose.

Parents who are very eager to find a diagnosis in their child may latch onto reports of a syndrome in which the features appear to match their own child's. However, as has been said, there is a huge overlap of features, and the overall picture of the child is critical. For example, a child may present with a combination of features such as hypotonia, abnormal head circumference, developmental delay and cleft lip or palate. There are a number of syndromes in which these features are seen, these include the four described below, with vastly different causes and recurrence risks.

- Trisomy 13, due to a chromosomal abnormality.
- Fetal valproate syndrome, due to the effects of the mother taking anticonvulsants during pregnancy.
- Crouzon syndrome, caused by a mutation in a single gene.
- Charge association, a combination of features often seen together but for which there is no single genetic basis known at present.

Luke Spencer is a 3-year-old with moderate learning problems. Luke was late in reaching all his developmental milestones, he started walking at 19 months, and was not regularly using two words together until 30 months of age. Luke is similar to his father in coloring, with fair hair and blue eyes, but he is on the 97th centile for height, whereas his mother and father are both relatively short in stature. Tracy, his mother, is not overly concerned about Luke, her brother is 'slow' and she says with disarming honesty that the teachers thought she was very 'thick' at school. Luke goes to a local playgroup but has difficulty communicating with other children and spends a lot of time playing alone.

When bathing Luke one night, Tracy notices his spine does not seem straight, and she rings the health visitor immediately. Luke is referred to a pediatrician, and is diagnosed with mild scoliosis. However, the pediatrician also notes Luke's learning delay and large head circumference, and takes a blood sample for chromosome analysis and Fragile X testing.

Luke is found to have an expansion of the FRAXA gene, which has caused his learning delay and probably the scoliosis. Tracy is tested and found to be a carrier of the expansion. She feels vindicated at her lack of achievement at school, and feels that with a lot of support she can help Luke to achieve his potential. Having a firm diagnosis for Luke helps the family to argue for a place in the opportunity playgroup for Luke, and he has regular speech therapy.

6. Genetic conditions in childhood

The conditions discussed here are chosen because they are commonly seen in a genetics clinic, because children with the conditions should be screened for complications, or because others in the family should be offered genetic counseling if the condition is diagnosed.

More information on each condition can be obtained from the references listed at the end of the chapter or via the websites listed in Appendix I.

6.1 Cystic fibrosis

Cystic fibrosis is a recessive condition caused by a fault in the *CFTR* gene [3]. This gene helps to control the flow of chloride ions through the cell membrane. As the balance of chloride membranes is abnormal, sodium and water balance is also affected. When a child inherits two faulty copies of this gene, the water/salt balance is altered in the mucus on the epithelial cells in the lungs and intestines. Recurrent respiratory infections occur due to the altered viscosity of the mucus, and the child will not be able to digest food efficiently due to the effect on the pancreatic enzymes.

The child with CF may have meconium ileus at birth, but recurrent chest infection, failure to thrive and loose, smelly stools are the most common signs in infancy. Although a blood sample may be taken from the child for DNA analysis, the diagnosis of CF is still usually made on the basis of an abnormal sweat test. Treatment includes daily physiotherapy to remove secretions, the addition of pancreatic enzymes to the diet, and prophylactic antibiotics.

If a child has CF, both parents will be carriers of the condition, and each of their other children will have a one in four chance of inheriting the condition. If the gene mutations are known, prenatal diagnosis is possible in each pregnancy.

There are a large number of potential mutations in the *CFTR* gene, and clinical testing is usually only carried out for 20–30 of the most common. Therefore, if a person has a CF carrier test that is negative, the chance that they are a carrier is greatly reduced, but there is still a small risk that they carry a rare mutation. The most common mutation found in people of Northern European origin is the delta F508 mutation, so-called because it involves a deletion at base number 508 in the *CFTR* gene [4].

Chester Family

Josie was 3 years old when her baby brother Ben was born. Ben had meconium ileus at birth, and was tested for cystic fibrosis at 6 weeks. His sweat test was abnormal, and cystic fibrosis was diagnosed. When speaking to the pediatrician, Josie's mother mentioned she had always been concerned about Josie's pale face and thinness, but had been reassured by her health visitor that some children are naturally pale and thin. A sweat test indicated Josie also had cystic fibrosis. Blood samples were taken for DNA confirmation of the diagnosis, and both children had two copies of the delta F508 mutation in the CFTR gene.

6.2 *Neurofibromatosis type 1*

Neurofibromatosis is an extremely variable condition, but because of the potential complications all children suspected of having NF should be referred to a pediatrician [5].

In many children, the appearance of multiple café-au-lait (CAL) patches on the skin will be the only sign. These benign coffee-colored marks usually appear between the ages of 1 and 5, and more than six CAL patches is diagnostic of NF. Children with the condition often also have a large head circumference and axillary freckling.

About 25% of children with NF have some type of learning difficulty, although this is often restricted to a particular area of learning, such as numeracy skills. Malignancies occur in a small percentage of children, and for this reason, unexplained signs and symptoms should be investigated promptly. As tumors can occur on the adrenal glands, blood pressure should be checked regularly. During adolescence and adulthood, those affected with NF usually develop some neurofibromata, small benign tumors of the nerve sheath.

These can cause pressure on nerves, but mainly cause difficulty due to the visual appearance. Individual neurofibromata can be surgically removed.

The condition is dominantly inherited, and often one parent will have mild signs of the condition. However, new mutations occur in the gene, so a child may be the first in the family to have the condition. If a parent has the condition, each child will be at 50% risk of inheriting it.

6.3 Duchenne muscular dystrophy

As an X-linked condition, Duchenne muscular dystrophy mainly affects males, although in very rare cases girls can be affected. The disease is caused by a mutation in the dystrophin gene on the X-chromosome, and may be carried by women who have no signs of muscular dystrophy.

This type of muscular dystrophy is generally diagnosed in boys when they are between 1 and 4 years of age. Generally the boy has been late in learning to walk, then is noted to have trouble climbing stairs or keeping up with his peers in terms of physical activity. The calves are usually large and firm. Greatly increased serum creatine kinase (CK) levels are a feature of the disease, and female carriers may also have raised CK levels. The diagnosis may be confirmed by DNA analysis or a muscle biopsy.

In the natural course of the disease, the large muscles are replaced by fatty tissue, and mobility is gradually reduced, with many boys requiring a wheelchair by the age of 12 years. Increasing deterioration of the heart muscle and restriction of breathing leads to death in the late teens or early twenties. A proportion of affected boys have learning problems [6].

When the diagnosis is made, the mother of the affected boy can be offered carrier testing. If a boy is the first member of the family to be diagnosed with Duchenne muscular dystrophy, his mother may be a carrier (2/3 of cases) or the mutation may have occurred for the first time in him (1/3 of cases). It is not always possible to identify the gene mutation in the affected boy. This can be a deletion, a missense mutation or a nonsense mutation. If direct mutation testing for the mother is not possible, linkage analysis may be used to try to clarify her carrier risk, particularly if she has daughters (who may be carriers) or if she wishes to have more children. Prenatal diagnosis is possible if the mutation is known or if linked markers are available to differentiate between the X-chromosome with the mutation and the normal X-chromosome.

6.4 Phenylketonuria

A recessive condition, for which all newborns in the UK are screened at 8 days of age (Guthrie test). Following a positive Guthrie test, a DNA test to confirm the presence of mutations in both copies of the phenylalanine hydroxylase (PAH) gene on chromosome 12 is performed. Children who are affected lack an enzyme needed to convert phenylalanine to tyrosine, hence the accumulation of phenylalanine in the body, and excretion of phenylketones in the urine. The increased phenylalanine levels damage the brain, and untreated children develop progressive severe mental retardation. Owing to the lack of tyrosine, there is little pigment in the hair and skin.

Treatment with a low phenylalanine diet reduces the brain damage, and many children have normal intelligence. However, recent studies have shown that the diet needs to be continued beyond adolescence or regression can occur. Women with PKU who are at risk of becoming pregnant should be advised to adhere to the low phenylalanine diet as the increased levels in their blood can damage the brain of the fetus. An NIH consensus statement in 2000 confirmed the need for multidisciplinary care for all patients with PKU [7].

6.5 Albinism

Albinism (lack of pigment) can occur as a result of a number of different genetic conditions. Some children have ocular albinism, a form that affects mainly the pigment of the eyes. This can have no clinical manifestations at all, but sometimes due to the lack of pigment the child will develop nystagmus, as the eyes try to focus.

Oculocutaneous albinism affects the hair, skin and eyes, and children with this form are at greatly increased risk of damage from the sun. Appropriate measures to protect their skin should be taken. In both forms, the child may benefit from dark glasses in sunlight.

As albinism condition can be inherited in a dominant, X-linked or recessive form, assessment by a genetic counselor should be advised if the family are seeking information on recurrence risks.

6.6 Down syndrome

Children with Down syndrome have a number of characteristic physical features, including short stature, broad neck, flat facial profile, upslanting palpebral fissures, inner epicanthic folds, small hands and feet, and large protruding tongue. About 40% of babies with Down syndrome also have a congenital heart defect. During childhood they are likely to require additional help with schooling, some attending mainstream school with assistance and others attending special schools for children with learning difficulties [2]. They may require surgical treatment that interferes with schooling, such as repair of hernias or congenital cardiac abnormalities. Eye surgery to 'normalize' the shape of the eyes is now being offered to some children, but remains controversial.

6.7 Turner syndrome

Girls with Turner syndrome are frequently diagnosed during childhood because of their small stature, or in adolescence because of the amenorrhea. The syndrome is caused by the absence of the second sex chromosome, girls with Turner syndrome having 45 chromosomes in all, and only one sex chromosome (45, X). This results in inadequate development of the ovaries, and most affected girls will not menstruate or ovulate. Infertility is therefore a feature of the condition.

Some girls with Turner syndrome do have learning problems, although the majority are able to deal with normal schooling. Rarely, a woman with Turner syndrome will have a child conceived using donor eggs. The syndrome usually occurs sporadically in a family, the recurrence risk for the parents is therefore low.

6.8 Sickle cell disease

Sickle cell disease is a recessive condition, the signs and symptoms of which are due to the altered shape of the red blood cells. Abnormal changes in the structure of the hemoglobin molecules cause the cells to form a sickle shape. They are more fragile, and block small blood vessels, resulting in both pain and anemia.

A child who inherits two copies of the faulty gene is said to have sickle cell disease. An individual with one normal and one faulty copy of the gene is said to have sickle cell trait. The presence of one faulty copy is known to help protect the individual against malaria, hence the high proportion of the population in some areas of Africa with sickle cell trait.

Sickle cell disease can be detected by a simple hemoglobin electrophoresis test.

Children with sickle cell disease are more prone to anemia, infection, delayed growth, and damage to kidneys or other internal organs due to blockage of blood vessels. In the neonatal period and early childhood they may be prescribed prophylactic antibiotics, and should be vaccinated against all childhood illnesses. Affected children should be encouraged to drink plenty of water, and folic acid may be given to reduce anemia. Details of current treatment for sickle cell disease are obtainable on the websites listed in Appendix I.

Carrier parents have a one in four risk of having a child with the condition, in each pregnancy. Prenatal diagnosis is possible.

6.9 Thalassemia

Thalassemia is a recessive disorder causing an abnormality of hemoglobin molecules. As the condition results in severe anemia for the affected child, the current treatment consists of regular blood transfusions (about every 4–6 weeks). The excess iron in the body that is released when red blood cells are destroyed accumulates in the liver and heart, and the child also requires administration of an iron-chelating agent (desferrioxamine) so that the excess iron can be excreted. Desferrioxamine is normally administered via a pump overnight. Some patients are now being successfully treated with a bone marrow transplant from a closely matched donor.

As there are two types of thalassemia, those affecting the alpha and beta hemoglobin chains, the exact diagnosis must be confirmed to enable the gene mutation to be found. If the mutation is not found, linkage studies can be used to track the faulty gene in the family, enabling prenatal diagnosis to be offered.

The website in Appendix I provides detailed information on the genetic aspects and treatment of thalassemia.

6.10 The autistic spectrum/Asperger's disease

A number of children referred to the genetics clinic will have Asperger's syndrome, or fit into the autistic spectrum of behavior. Asperger's syndrome is classified under the *Diagnostic and Statistical Classification of Mental Disorders* (DSM-IV), and is defined as a condition in which there is little verbal or cognitive delay, but children with the condition have marked lack of social skills, in particular finding it difficult to interpret nonverbal communication. They are also prone to repetitive or obsessive patterns of behavior. Although there are a number of studies currently being conducted into the potential genetic cause of **autism**, other theories include the idea that autism is caused by an immunogenetic susceptibility to pathogens during pregnancy that affect a particular fetus but may not be harmful to others.

At present, if a child does not have significant dysmorphic features or other disabilities, it is unlikely that a genetic cause will be found to explain Asperger's syndrome. Recurrence risks for the parents of a single child with Asperger's are low, and as a specific gene mutation has not been identified, prenatal diagnosis is not possible.

6.11 Fragile X syndrome

Fragile X syndrome (FRAX) is the second most common cause of learning delay in boys (the most common is Down syndrome). The syndrome was named because of the fragile sites seen on the X chromosome when the cells of an affected person are specially treated before culturing. It is an X-linked condition, and female members of the family may be carriers. Boys with the condition have moderate–serious learning problems, usually requiring special schooling. They are often tall in stature, and have large ears and testes.

The syndrome is caused by an expansion in a particular gene on the X chromosome, therefore a definitive diagnostic test is possible, and female members of the family can be tested for carrier status if they wish.

6.12 Hearing impairment

Sensori-neural deafness is most often detected in children in infancy or early childhood. If both parents are hearing, then the condition is likely to be recessive, and future children born to the couple will have a one in four risk of hearing impairment. However, some forms of deafness are dominantly inherited. A careful family history is needed before recurrence risks can be given. Obviously, cases of conductive deafness due to infection, accidents or aging are not relevant to the genetic issues.

Some parents who have a hearing impairment themselves feel that the ability to communicate does not depend upon hearing, and do not therefore consider deafness to be a disability.

7. Genetic testing of children

Genetic testing of children has already been discussed in *Chapter 1*, however, it is relevant to revisit the topic here in more detail.

Guidelines for testing children were suggested by the Clinical Genetics Society in 1994 [8], based on studies performed at that time, however, there has been little empirical work to demonstrate whether individuals are actually harmed by testing in childhood for adult-onset diseases. As there is little evidence either way, it seems prudent to adopt the approach of doing least harm.

In general, if children are at risk of a condition that would require treatment or surveillance in childhood, then testing is justified. In some cases, testing a child to establish a diagnosis would not harm the child, and would possibly be of great benefit to the family.

However, if a child is at risk of a condition that normally only affects adults, then testing is better delayed until the child can give informed consent. There is some evidence that when an individual is part of the decision-making process for testing, then they are better equipped to deal with the results of those tests. Testing children for carrier status would also fit into this category, as the results would only be of relevance to the individual when having children.

There are circumstances that arise in which withholding a test for a child may appear to damage the family unit, in those cases, consideration is given to the particular case, and expert advice is usually sought from other professionals in the field of genetics and related disciplines such as medical ethics. The establishment of clinical ethics teams in healthcare settings will be of great assistance in these cases.

7.1 *Testing children – two case comparisons*

Collins Family

Carol Dixon (nee Collins) is the mother of three children, Adam (11 years) Joe (8 years) and Amy (4 years). Carol has a family history of colon cancer, and had a colectomy at the age of 29 years, when she was found to have a bowel obstruction shortly after the birth of Amy. She found the diagnosis of colorectal cancer very difficult to come to terms with at first, but is now very positive about the surgery 'that saved her life.'

Carol was told that she had hundreds of polyps in her bowel, and that her children will need to be screened from the age of puberty.

She sees the genetic counselor, who explains the inheritance pattern (dominant) and offers to take a blood sample from Carol to see if the gene mutation for polyposis coli can be identified in the family. If the mutation is found, presymptomatic testing will be possible for the children. Carol is very keen, and a sample is taken. The counselor explains that it may be months or even years before the mutation is found. If the mutation is not found within a year, colonoscopy for Adam may have to be considered.

Four months later, the results arrive, the mutation has been found in Carol's sample, and she meets the counselor again to discuss testing for the children.

The counselor is aware that Carol very much hopes that the test will show that the children are 'all clear.' When the counselor asks her for her reasons for wanting the children tested, she says it is so she 'doesn't have to worry about them any more.' The counselor spends a lot of time encouraging her to consider the possibility of a positive result, that is one that confirms the mutation is present in one, two or all of the children. One of the ways the counselor does this is to rehearse the news-giving session with Carol, using the terms that would be used for either outcome of testing. It is when the counselor rehearses saying 'I'm sorry but Adam does have the gene mutation and will develop polyps in the bowel' that Carol starts to realize that the outcome of testing could go either way, and breaks down. They talk about her guilt at possibly having passed the mutation on, and how desperately she wants to protect her children from worry and ill-health.

At the next session, the counselor and Carol talk about the children's readiness for testing. Carol feels she would like to know the status of all three children immediately. However, when they discuss how much understanding each of the children has about the situation, she agrees that only Adam would really understand the test and the reason for it.

At two further appointments, Adam and Carol are seen together, and blood is taken from Adam for a test, at his request. Four weeks later the counselor meets Adam and Carol to give them the news that Adam has inherited the gene mutation and colonoscopic screening is recommended. Adam has a colonoscopy 4 weeks later, several polyps are treated with laser therapy and Adam is scheduled for colonoscopy on an annual basis.

The counselor arranges to meet Carol and Joe in 2 years to discuss testing for him.

Peter Harding is 42 years of age. He has Huntington's disease. He and his wife Angie have separated, but still see a lot of each other. Angie helps Peter with his household chores, and has him to her place for a meal every second day, as he finds cooking for himself a bit of a problem, and he likes to spend time with their two children, Melissa and Jason. They all get along better since the divorce, there aren't so many arguments and now Angie knows what is causing Peter's moods, she is able to deal with them better.

Angie is very anxious about Melissa and Jason. She knows they are at 50% risk, and wants them to be tested. When she meets the genetic counselor to discuss testing, she says that she is worried Melissa might get pregnant, but if she knows the baby is at risk of HD that might make her more careful.

The counselor explains that many people at risk of HD do not want to know their status, preferring instead to live with the risk. She also explains that informed consent is needed for such a test, and that children are not tested except in exceptional circumstances.

Angie brings Melissa to the next appointment, and it is clear that Melissa does not want to be tested at present. She is very upset about her father's illness, and is trying to put her own risk out of her mind and 'just get on with life.' Angie asks the counselor to take a sample from Jason without telling him what it is for, because he is a worrier. Again, the counselor explains that testing without informed consent is not ethical.

The genetic counselor offers Angie several further appointments to discuss her own anxieties for her children. She leaves the door open for Angie to contact her again if she wants further counseling, and tells the children they can see her again to discuss their risks in the future if they wish.

8. Genetic healthcare issues in adolescence

8.1 Transfer of care from pediatric to adult services

A young person who is affected with a genetic condition such as cystic fibrosis or muscular dystrophy will have healthcare provision under the pediatric services until at least the age of 16 years. In reality, health care for these clients usually continues to be provided by the pediatrician beyond that age, partly because of the long-term relationship that has been built up with the young person and their carers. However, sometimes this occurs because adult services for these clients are not as well developed as pediatric services. This is in part due to the changing survival rate; adult services for such clients were not in demand when most children died from the condition. In other situations, children are discharged from pediatric care but there is a noticeable gap in health care for them in adult services.

Those caring for young people should, however, bear in mind their emotional and psychological needs as well as the physical needs, and it may not be very positive for a young person of 19 or 20 years to be cared for in a pediatric clinic. Adolescence is normally a period in which young people establish their independence, gradually loosening the ties to their parents. Disability brings with it additional dependence on others; this is reinforced by having to use pediatric facilities.

In some areas, a joint pediatric/adult clinic is set up to expedite the effective transfer of care, and if this can be organized, it may be very helpful in insuring the young person's physical and psychological needs are met.

8.2 Issues of sexuality and reproduction

Experience working with disabled teenagers has shown that they think about their own sexuality and reproductive issues far earlier than their parents realize! Evidence suggests that young people often avoid talking about such issues with their parents to protect them, but that these issues are very important to them. It is therefore helpful to be able to offer young people affected with a genetic condition opportunities to talk freely about their condition and ask questions. A school nurse, practice nurse, family doctor or genetic counselor can offer this, but sometimes it is possible to set up contacts between special schools and genetic counselors, so that students can be offered the chance to speak privately to the counselor.

Although young people may ask about the chances of their own children developing the same condition, they are frequently concerned about their own future prognosis, and wish to speak with someone who has experience of the condition.

8.3 Informing the adolescent about genetic risk

One question that is often asked by parents who have children at risk of a genetic disease is 'When should I tell my child about the risk?'. Parents frequently feel the conflict between believing that a child has the right to know about the risk, and wishing to protect the child from worry [9].

Studies show that those who are at risk wish to be told, mainly so that they have the opportunity to make life choices, taking the risk into account. There is no 'right time' to tell a child, but it appears that individuals who grow up knowing about the condition in the family are able to adjust and live with the risk more easily than those who are told in adulthood. Those at risk find it especially hard to adjust if others in the family knew of the condition, and withhold the information from them. This secrecy, that exists in many families at risk of a genetic condition, creates anger and distrust, and hard as it is for parents, it seems much more positive to tell children of their risk and so empower them to consider the options available and make their own decisions.

A study focusing on telling adolescents about their risk of polyposis coli showed that it was better to avoid telling the child in the period of early adolescence. Children who were told either before 10 years or after 13 years were able to accept the news more easily than those who learnt of their status during the critical period between those ages. As so much change occurs at that time, when the child is struggling to develop a sense of their independent self, this is not surprising.

CASE EXAMPLE JOSEPH

Joseph was the eldest of two children born to Harry and Helen. Harry's father died of presenile dementia at 48 years of age. At that time the family were unaware that this could be an inherited condition. Harry's much older sister developed the condition in her 40s, it was then that the family began to suspect it could be familial. However, Helen and Harry both wanted children, so went ahead with their family plans.

Harry was diagnosed when Joseph was 17 years of age. His young sister was 14. Joseph suspected the condition was familial, but his mother denied this vehemently. Finally, Joseph sought medical advice prior to his marriage. Joseph found out that his father's condition was an inherited form of presenile dementia and that he was at 50% risk of developing it himself. He was furious at having been lied to, and an argument with his mother followed. Joseph's mother's rationale for not telling her children was that she didn't want them to worry.

Joseph felt his sister should also be aware of her risk. His mother warned him that if he told his sister he would be cut off from the rest of the family. As he was very concerned about his father he wanted to avoid this. He felt forced to comply.

Without the knowledge of her family, Joseph's sister found out about her risk. She is now 20 years old and wants a family. She is terrified of having a child because of the risk, but feels unable to talk to the family about it. Because of the secrecy, both Joseph and his sister feel alone and unsupported by the family.

8.4 *Loss of a parent in adolescence*

In a genetics clinic, there will be a considerable proportion of clients who have lost a parent during their childhood or adolescence. This is especially true of clients seeking advice about a family history of cancer, and this history may have a strong influence on decision-making. For example, a woman who was in her teens when she lost her own mother because of breast cancer will have been influenced by that event. Her attitude to her own health, her body, her sexuality and her relationships will have been affected, and this may alter radically her approach to genetic testing, screening programs, breast self-examination and prophylactic surgery. In particular, women who lost a parent before they were independent frequently express a need 'not to leave my children without a mother,' and this may influence them to opt for radical prophylactic surgery.

9. Conclusion

In childhood and adolescence, genetic conditions may cause a range of physical and/or learning problems. These will have an impact on the child's social and educational development, and healthcare services for these children must address these areas as well as the treatment or prevention of illness. As the child grows into adolescence they may face particular difficulties in establishing independence, and sensitivity is needed to enable the young person to establish independence at an appropriate level.

TEST YOURSELF – DISCUSSION CASE

Joe is 17 years old, and has learning problems. He attends a special school, and prefers to play with children who are about 3 or 4 years younger than himself. His reading age is 9 years.

The cause of Joe's problems has not been investigated. It is suspected that he may have fragile X. His sister Marie is married, and worried that she may have a child with learning problems.

A blood sample from Joe could be tested for fragile X and chromosome abnormalities. The results of these tests would help the counselor advise Marie about the risks of her children having learning problems.

Joe is reluctant to be give a blood sample, he hates 'needles'!

What are the rights of Marie and Joe in this instance? Should Joe be persuaded or even coerced into giving a sample? Do Marie's future children have any rights in this matter?

Would your thinking change if Joe was only 5 years old?

Discuss with colleagues what you would do if (a) Joe was your patient (b) Marie was your patient. Is there an ethical solution?

References

1. Skirton H (2001) The client's perspective of genetic counselling – a grounded theory approach. *J Genet Couns* **10**: 311–330.
2. Jones K (1997) *Smith's Recognizable Patterns of Human Malformation*, 5th edn. Philadelphia: W.B. Saunders.
3. Cheng SH, Gregory RJ, Marshall J *et al.* (1990) Defective intracellular transport and processing of CFTR is the molecular basis of most cystic fibrosis. *Cell* **63** (4): 827–834.
4. Kerem B, Rommens JM, Buchanan JA *et al.* (1989) Identification of the cystic fibrosis gene: genetic analysis. *Science* **245** (4922): 1073–1080.

5. Gutmann DH, Aylsworth A, Carey JC et al. (1997) The diagnostic evaluation and multidisciplinary management of neurofibromatosis 1 and neurofibromatosis 2. *JAMA* **278** (1): 51–57.

6. Emery AE. Clinical and molecular studies in Duchenne muscular dystrophy. *Prog Clin Biol Res* 1989; **306**: 15–28.

7. NIH (2000) Phenylketonuria (PKU): screening and management. *NIH Consensus Statement* **17** (3): 1–33.

8. Clinical Genetics Society (1994) *The Genetic Testing of Children*. Clinical Genetics Society.

9. Skirton H (1998) Telling the children. In AJ Clarke, editor. *The Genetic Testing of Children*. Oxford: BIOS.

Further reading

Clarke AJ (1998) *The Genetic Testing of Children*. Oxford: BIOS.

Jones KL (1997) *Smith's Recognizable Patterns of Human Malformation*, 5th edn. Philadelphia: W.B.Saunders. (Seminal text on dysmorphic syndromes.)

9 Adulthood

1. Introduction

This chapter is concerned mainly with those conditions that have an effect on the health of a person during their adult life. These are usually termed the adult-onset conditions, but because of the variability of some conditions, there may be a huge range in the age of onset, and even within the same family the onset of signs and symptoms may occur at vastly different ages.

Although there are many adult-onset conditions, for simplicity in this chapter we will focus mainly on four groups of conditions; familial cancers, neuro-muscular disorders, psychiatric disorders and genetic hemochromatosis.

2. Familial cancer

2.1 *Genetic basis of cancer*

Owing to the high incidence of cancer in the population, there will be some history of cancer in virtually every family. However, in the majority of cases cancer occurs as a sporadic event, due to changes in the genes in a particular cell (somatic change). With the media coverage of familial cancer, and the increasing awareness of the genetic influence on a number of conditions, more people are wondering if they have an increased susceptibility to cancer, and if so, how can they increase their chances of survival. Although there are a number of rare cancer syndromes, such as von Hippel Lindau disease, in general the majority of familial cancer referrals will be connected with colorectal cancer or breast/ovarian cancer. If a person presents with a family history of cancer, we need to try to work out if a number of cancers have occurred co-incidentally in the family, or if there is a family gene mutation being inherited by some family members.

The function of some genes is to help prevent the growth of tumors or cancers, these are called tumor suppressor genes. Whenever new cells are produced in the body, to replace dead or damaged cells, the tumor suppressor gene limits the number of new cells, thus preventing overgrowth of the tissue. If the particular sequence of these genes is correct, then the protective action of the genes is intact. However, the genetic material is recopied over and over again during the person's lifetime, as new cells are made. Each time the gene is copied, there is the potential for a mistake to be made. If one copy of the gene becomes faulty or is 'spelt incorrectly' then usually the remaining normal copy will still protect against cancer. However, if both copies of that particular gene become faulty then the protection against cancer will be removed, and the person will usually develop a tumor [1].

Knudson [2] first described the two-hit hypothesis in development of cancer in retinoblastoma, and the hypothesis helps us to make sense of what occurs in familial cancer (*Figure 1*). If a person is born with one faulty copy of a gene, then they are likely to develop cancer at a younger age, because they only need one more accidental fault in the gene to occur to make both copies faulty.

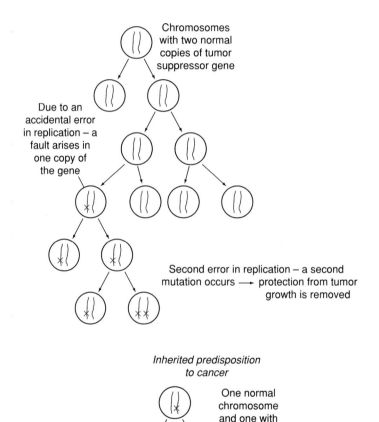

Figure 1. Two-hit theory.

If you are interested in knowing more about this theory, there is a website address in Appendix I.

When we look at a family history, there are three main indicators that a gene mutation may be causing a predisposition to cancer in the family.
- Are the cancers in the same or related parts of the body?
- Is there cancer in more than one generation of the family?
- Have people in the family developed cancer at a younger age than you would generally expect?

Take for example the cases of two women referred with a family history of breast cancer, Gail and Helen.

In both cases, the referral letter gave the same information.

> Dear counselor
>
> This woman is 38 years of age. Her mother had breast cancer, and her grandmother also died of cancer.
>
> She is very worried about her own risk. Can you assess and give me some guidance about screening for her?
>
> GP

The counselor sees both women, and takes a family history.

Gail: In Gail's case, her mother had breast cancer diagnosed last year, when she was 63 years old. The grandmother who died had cancer of the bladder at 72 years of age (*Figure 2*). Gail is at low risk because although there are two generations affected with cancer, cancer of the bladder is not known to be caused by the same gene mutation as cancer of the breast, and both cases occurred in older age, making them more likely to be due to the aging process rather than an inherited genetic predisposition.

Additional screening for Gail is not indicated.

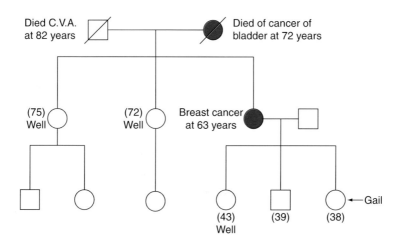

Figure 2. Gail's pedigree.

Helen (*Figure 3*): Helen's mother was diagnosed with breast cancer at the age of 38 years. Although the treatment at the time was thought to be successful, she subsequently died after developing metastases 3 years later. Helen was only 15 years old at the time of her mother's death.

Helen's maternal grandmother, Hilda, had 'stomach' cancer at 58 years and died within 2 weeks of diagnosis. A check with the cancer registry showed this was in fact ovarian cancer.

Breast and ovarian cancer are known to be caused by the same gene mutation, Helen's mother was very young when the cancer was diagnosed, and there are two successive generations affected. It is likely that in this family there is a gene mutation that increases susceptibility to breast and ovarian cancer. Helen is at high risk, and should be offered both breast and ovarian cancer screening.

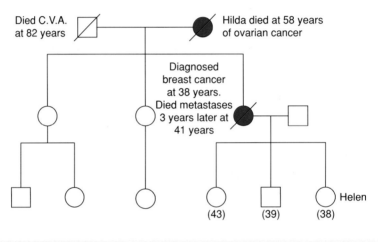

Figure 3. Helen's pedigree.

2.2 *Breast and ovarian cancer*

Guidelines for referral to the genetic service. Health professionals working in primary or secondary care will often be faced by a client who has questions about their own family history of cancer. It is helpful to have some guidelines as to whether a history is likely to be significant.

If a person's family history fits within any of those listed below then this may indicate a higher risk of developing an inherited form of breast and/or ovarian cancer and a more detailed assessment should be made. NB *A close relative means a parent, brother, sister, child, grandparent, aunt uncle, nephew or niece.*

Many families are not aware that inherited forms of breast and ovarian cancer can also be passed down through the father's side of the family, therefore, the family history from both sides of the family could be significant. Men can carry the faulty gene, but of course are less likely to develop this type of cancer since they have no ovarian and very little breast tissue. Women who are

BREAST CANCER HISTORY – GUIDELINES FOR REFERRAL TO GENETICS SERVICE

Three close relatives from the same side of the family diagnosed at any age

OR

two close relatives from the same side of the family with an average age of diagnosis under 60 years

OR

mother or sister diagnosed under 40 years

OR

father or brother with breast cancer diagnosed under 60 years

OR

one close relative with bilateral breast cancer, with the first cancer diagnosed under 50 years.

Breast and ovarian cancer

One close relative diagnosed with ovarian cancer at any age and at least two close relatives with breast cancer with an average age of diagnosis under 60 years, all from the same side of the family

OR

one close relative diagnosed with ovarian cancer at any age and at least one close relative diagnosed with breast cancer under 50 years from the same side of the family

OR

one close relative diagnosed with breast cancer under 50 years and ovarian cancer at any age.

Ovarian cancer

Two close relatives from the same side of the family, at least one of whom is either a mother or sister, diagnosed at any age with ovarian cancer.

of Ashkenazi Jewish origin are at higher risk, due to the presence of particular gene mutations in that population, and should be assessed even if the history does not meet the criteria stated above.

Tables developed by researchers such as Claus *et al.* [3] can be used to determine the level of risk to each person in the family, and as a guideline for screening recommendations. Software packages such as *Cyrillic* are based on these tables and can also be used for this purpose. However, a level of expertise and underlying knowledge is necessary in order to interpret the family history data correctly and so use the tables accurately. A publication by Eccles *et al.* [4] provides a useful tool for interpreting family history of breast and ovarian cancer, which is simpler to use.

Screening protocols. At present, there are no ideal methods for screening women for breast cancer. Women are advised to examine their own breasts after a period each month, and to seek medical advice if they see or feel changes in the breast or armpit.

Mammography is used extensively, and at present is the screening method of choice, but is less sensitive in premenopausal women than in those who are postmenopausal, and so is less effective in the high-risk group for whom additional screening is needed. However, evidence that screening is worthwhile for women less than 50 years has emerged [5]. Trials of ultrasound screening are ongoing.

Similarly, ovarian screening is not highly sensitive. Ultrasound of the ovaries is used, but may not detect a tumor. The CA125 test, which is a measurement of a hormone excreted when there is a malignancy in the ovary, may also be used, and if there is a suspicion of a tumor on ultrasound or CA125 testing, an ovarian biopsy will usually be performed.

2.3 Colorectal cancer

When there is a history of colorectal cancer, we can categorize families into two levels of genetic risk, according to the amount of evidence supporting the presence of a gene mutation in the family. The first group of families are those where it appears very likely there is a dominant gene involved.

Familial adenomatous polyposis (FAP) – also called polyposis coli. Owing to a change in the *APC* gene, multiple polyps grow inside the colon. These predispose to adenocarcinoma of the colon or rectum. If a person has the gene mutation, they are almost certain to develop dozens or even hundreds of polyps in their teens or 20s. Family members who are at risk should be screened annually by colonoscopy, from early adolescence (10–12 years) [6]. If a person develops multiple polyps, colectomy is performed to reduce the cancer risk, often in the late teens. Every child of an affected person has a 50% chance of inheriting the condition.

Hereditary nonpolyposis colon cancer. If there is a strong family history of bowel cancer, but the affected individuals have not had multiple polyps, then one of the genes known to cause hereditary nonpolyposis colon cancer (HNPCC) is suspected. In addition to colorectal cancer, these gene mutations may also be implicated in endometrial, stomach, ovarian and urinary tract cancer.

The genes involved are **mismatch repair genes** (such as *MLH2* and *MSH2*), that is they have a role in repairing faults in the DNA sequence. The mismatch results in microsatellite instability, so initially a tumor sample can be tested for instability to try and locate the site of the gene mutation.

Regular screening by colonoscopy from about 25 years of age is recommended if a person is at risk of HNPCC. Current guidelines for management of at risk individuals can be accessed from the website of the ICG (see Appendix I for address).

Colorectal cancer predisposition. There are a number of families where there is *some* evidence for the presence of a gene mutation, but the cancers in the family may also have occurred sporadically. In these families, screening may be undertaken on a less frequent basis, for example every 5 years, starting 5 years before the mean age of diagnosis of colorectal cancer (CRC) in that family.

Lovett [7] developed some guidelines that help to clarify the risk when there is a family history of colorectal cancer (without multiple polyps).

Family history	Lifetime risk
> 2 first-degree relatives affected	1 in 3
2 first-degree relatives affected	1 in 6
1 first-degree relative < 45 years affected	1 in 10
1 first-degree relative AND 1 second-degree relative affected	1 in 12
1 first-degree relative > 45 years affected	1 in 17
General population risk of colorectal cancer	1 in 35

In general, the following current guidelines for screening are used. The frequency varies according to the level of risk.

History	Screening starts	Frequency
FAP	12–13 years	Yearly
HNPCC	25 years	≤ 2 yearly
2 first-degree relatives affected 1 first-degree relative < 45 years 1 first-degree relative + 2 others same side	Start at 5 years younger than mean age of cases	5 yearly (3 yearly if polyps seen)

2.4 Genetic counseling for familial cancer risk

When a family is referred because of a perceived risk of familial cancer, it is extremely important to document the family history accurately and confirm the diagnoses of cancer whenever possible.

Although consent is not required to search the records of individuals who have died, it is required from family members who are still living. In the initial stage of the contact, forms may be given to the referred person to help them obtain the relevant information and to document the written consent from their relatives.

The network of cancer registries in the UK is useful in obtaining confirmation of cancer diagnoses, but in some cases medical records or death certificates are used.

After confirmation of as many cases as possible, the risk to the referred person can be discussed, and recommendations for screening made.

2.5 Presymptomatic genetic testing in familial cancer

In both familial breast and bowel cancer, there may be a number of potential genes involved, and the mutation may differ from family to family. It is therefore necessary to identify the faulty gene in each family before genetic testing can be offered to unaffected family members. Samples from an affected person are required for analysis, and only when the gene mutation is found can presymptomatic testing be offered to others in the family.

Laboratory science has not yet progressed to the point where the mutation can be found in every family, therefore some individuals who would like to know their status are unable to be tested.

If a person does request testing, and the mutation is known in the family, then the counselor works with the client to help him or her to prepare for the result. The preparatory counseling usually occupies 2–3 sessions at least.

Following the result, clients are generally followed up for some time, whatever the outcome. Those who are shown to have inherited the mutation will of course be advised to continue with their screening program. In some cases, clients who know they carry the gene mutation may opt for prophylactic surgery, such as mastectomy or oophorectomy. This is especially true of women who have experienced the loss of close relatives with breast and or ovarian cancer, and who may be very concerned about leaving their own families without a mother.

There is evidence that women are more likely to communicate about health matters within a family, and to take the responsibility for informing others of risk. When a mother dies, her children may lose the person who would have been their main informant on these matters, although grandmothers and aunts often take over this role. For health professionals, it is helpful to be aware of the potential gap in health-related information that may exist in these families.

2.6 Psychological implications of risk status

One of the roles of the health professional is to encourage clients who are at high risk to use the means available to increase their chances of survival, whether this is by self-examination, self-reporting or undergoing clinical investigation. Providing education on effective techniques for breast self-examination, or information on the potential signs and symptoms of bowel cancer is well within the remit of the primary care team or nurses working in secondary care, for example in a surgical team.

There are many people at increased risk of cancer who seek clinical screening and who find reassurance in the knowledge that they are doing all they can to detect early cancer. However, in some cases the fear of discovering a tumor is so great that it impedes the person's ability to comply with screening. Lynch and Lynch [8] described this scenario in families at risk of bowel cancer, where

Collins Family

Carol Collins was diagnosed with colorectal cancer, and had a total colectomy when she was 29 years old. The histology report indicated multiple polyps in the section of bowel examined. The presence of hundreds of polyps in her bowel is diagnostic of FAP. Carol's mother Peggy died at 44 years with colon cancer, and it is likely she also had FAP, although the histology records are not available and the cancer registry records only the diagnosis of colon cancer. Carol says her mother was 'too far gone' for treatment when she was diagnosed.

As it is likely that the diagnosis in this family is FAP, those at risk in the family should be offered colonoscopic screening. Adam, Carol's oldest son is now 11 years of age. There are two other children in the family, Joe who is 8 years, and Amy, 4.

Carol's brother Robert (22 years) has been suffering some rectal bleeding. Another sister, Miriam is only 18 years old and has never been screened.

A blood sample is taken from Carol for genetic testing. Initially, a mutation is not found, so testing cannot be offered to Miriam or Robert. However, as they are at risk, they are urged to consider colonoscopic screening. Robert is reluctant, but finally agrees after pressure from his girlfriend. He does not turn up for the screening appointment.

Miriam has a colonoscopy, and no polyps are found, this indicates she is unlikely to have inherited the gene mutation as we would have expected some polyps to have grown by the age of 18 years. She is, however, asked to return for screening again in 2 years.

Robert is encouraged by her result, and goes for screening. He has multiple polyps and a small malignancy in the colon. An emergency colectomy is performed immediately, but further therapy is not needed.

Several months later, the gene mutation is found in Carol's sample, and her brother and sister are tested. Miriam's test confirms she did not inherit the mutation and she requires no further screening. Miriam is both relieved and guilty about having 'escaped' the family condition. Robert has inherited the mutation, and his two children are at 50% risk themselves. This creates another challenge within the family, as Robert has lost touch with their mother and is unable to inform her that their children are at risk.

fear is overwhelming, and the client therefore will not attend for colonoscopy. This is certainly evident in clinical practice. There are also many women at high risk of breast cancer who do not examine their own breasts regularly, because of the fear of discovering a lump.

2.7 *The role of intrusion and avoidance*

The psychological concepts of intrusion and avoidance [9] are helpful in understanding responses to risk. Some women report examining their breasts regularly for a period after they are reminded of their risk, for example after seeing the genetic counselor or after another member of the family has a 'scare'. At those times, the cancer risk 'intrudes' into the thoughts frequently. However, intrusion is difficult to maintain, and the mind's response to constant intrusion is avoidance. Gradually, when the intrusion lessens, and the risk is less prominent, the motivation to self-examine is reduced.

The motivation to self-examine may be increased if the woman has encouragement from the primary care nurse, especially if the nurse is able to allocate time to meet with the woman regularly, perhaps once or twice a year, to check the breasts and discuss any concerns.

When a client continually fails to attend for screening, the reasons for such nonattendance may be complex. This is especially so if the client confirms the need for screening, makes appointments, but does not attend. As stated it may be fear of a tumor being discovered that prevents attendance, but there may also be other reasons.

> ### Collins Family
>
> *Robert is at risk of FAP, and has rectal bleeding. He agrees to have a colonoscopy but does not turn up for the appointment. Robert has met the genetic counselor several times, and she visits him at home a week later. He is reluctant to say why he did not attend. After a long conversation, Robert finally admits that he has spent time in prison, during which he was raped by a male inmate. His fear of colonoscopy stems from this experience of anal rape.*

3. Huntington's disease – a model for predictive testing

3.1 Description of Huntington's disease

Huntington's disease (HD) is a condition that affects the physical, mental, emotional and social health of the affected person. It has been known since George Huntington first described the condition in the literature in 1872 [10] that the disease is dominantly inherited, but until the mutation was identified in 1993 the concept of '**anticipation**' was puzzling. 'Anticipation' refers to the phenomenon in which succeeding generations of the family develop the disease at a younger age than the preceding generation. In families affected by HD, anticipation sometimes occurs, but sometimes not.

The mutation in the *huntingtin* gene is an expansion, that is, abnormal genes are longer than the normal gene. In the genetic material, certain base pair sequences are often repeated within a gene. Within the *huntingtin* gene, unaffected individuals have up to 35 repeated copies of the sequence 'CAG'. However, affected patients have more than 35 copies of the CAG triplet in one copy of the gene. They thus have one normal and one expanded copy of the gene [11].

The CAG **trinucleotide** codes for glutamine, and the expanded gene increases the length of a glutamine chain in the cell cytoplasm, causing it to form clumps and invade the nucleus of the cell. This results in premature death of brain cells.

Once the number of CAG repeats in the gene has expanded, it is less stable and the number of repeats can increase when the gene is copied during meiosis. The expansion is more likely to increase during spermatogenesis than oogenesis. Hence, if a man has HD, his children may inherit a larger copy of the gene than he has, and develop the condition at an earlier age than he did.

The signs and symptoms of HD vary with each individual, but fall into three main groups:

1. *Physical disability*: The first physical signs may be clumsiness, stumbling and unsteadiness. Clients often see a deterioration in their hand writing, may trip when walking, and may spill drinks or food more frequently than usual. Eventually, chorea may be evident, and walking becomes very difficult. The speech becomes slurred.

2. *Mental disability*: Initially, the client is often aware that their memory is becoming worse, especially short-term memory. Ability to do mental arithmetic can also deteriorate, this can be evident when shopping. The 'executive functions' suffer, and clients will often find it hard to problem-solve or respond to changes in plans or situations. Some clients eventually suffer serious dementia.

3. *Psychiatric problems*: Depression may be the first sign of HD, and can be effectively treated with antidepressants. Paranoia and obsessive behavior may also be present; these are naturally difficult for the family to deal with, but can also be treated by the psychiatric team. Suicide is a significant risk.

The age of onset is related to the size of the expanded fragment in the gene, and varies from childhood (rare) to the eighth or even ninth decade. However, the majority of people are diagnosed in their 40s.

3.2 Predictive testing

For some people at risk, the uncertainty of their situation is difficult to bear, and they choose to have a predictive test. This type of test is performed on a sample from a healthy individual, to determine whether the person has inherited the gene mutation that will eventually cause the signs and symptoms of HD.

Prior to the test being available, studies showed that the majority of those at risk would request a test, but these results are not borne out by the actual uptake of predictive testing. However, fewer than a quarter of those at risk actually request testing. This is understandable given that there are no preventative measures that could delay the onset of the disease, no cure and

Mary is 50 years old, and the daughter of Cyril Harding. Cyril was diagnosed with Huntington's disease when he was 50 years of age. He had been with the same firm since leaving school, working as an electrician and though he was unable to continue his normal work, they found a less demanding job for him in the stores department. He managed to work for 5 years after diagnosis, then took early retirement. He died at 62 years of age of pneumonia after breaking his hip in a fall.

Mary started showing some signs of HD in her late 30s, although it wasn't diagnosed until she was 41 years old. At that time she had a CT brain scan that showed significant changes.

Mary gave up work at 43 years of age. She said it was all becoming too high-powered at the garden center where she worked. Everyone had to learn to use a new barcode system and she just couldn't get the hang of it, anyway, her husband made enough to keep them comfortable, especially once the kids had left home.

Mary has never had a genetic test, never saw the need for it, she knew she was just like Dad. It didn't seem fair though, that she got it so much earlier than him.

no effective treatment. Most people prefer to retain some hope that they have not inherited the condition.

A protocol for predictive testing has been in use in the UK and many genetic centers worldwide since testing commenced [12,13]. This protocol is aimed at insuring that clients who are tested have adequate opportunity to explore the implications for themselves and their families, and to prepare for either result. The test is only performed with informed consent.

Some of those who come forward for discussion about testing do so at the instigation of their family or health carers. However, the aim of the preparatory discussions is to help the client decide whether certainty is preferable to uncertainty, even if they receive 'bad news.'

For those who do decide that certainty is preferable, several counseling sessions are offered to help the client prepare for the result. The client is encouraged to bring a support person to the sessions, and is helped to plan the period following the results. Interestingly, those who have lived with the risk of HD for a number of years often find it difficult to adapt psychologically to not being at risk, as this requires a greater adjustment than finding out they are going to be affected. Many also suffer from survivor guilt, particularly if their siblings are affected. Strong emotional support is often required after the test, whatever the outcome, and when the result is good this is more often lacking as friends and family cannot understand why the client is having difficulties.

The predictive testing protocol developed for use in HD testing is also used for predictive testing in other genetic conditions, such as CADASIL, myotonic dystrophy, or adult polycystic kidney disease, and HNPCC. It is a

Harding Family

Sarah is Mary's daughter, and the granddaughter of Cyril Harding. Sarah is now 22 years old, and has known about her risk of HD since her mother was diagnosed, about 9 years ago. It was a great shock to learn about the family condition at that time, and Sarah has broken off several relationships rather than tell a boyfriend about her risk.

Now Sarah has met a man she really loves, and told him about the HD after he met her family. Naturally, he could see her mother was ill, as she uses a wheelchair and he had trouble understanding her speech.

Sarah and Tom plan to get married, and would like a family. They come to see the genetic counselor to discuss HD, and the risks to their future children. Together they decide that if Sarah has the gene they will get married, but will not have any children.

Before testing Sarah, the counselor visits Mary to ask if she can have a sample of blood to confirm that HD is the correct diagnosis. Mary enjoys the attention, and gives her consent for her sample to be tested. The test shows that Mary has one copy of the gene with 19 CAG repeats, and one copy with 45 repeats.

After three counseling sessions, Sarah has blood taken. Two weeks later she and Tom meet the counselor for the result to be given in person. Sadly, Sarah has inherited HD from her mother. She and Tom are distraught, but feel they have the information they need to plan their lives together with more certainty.

Three months later Sarah and Tom marry. It is a happy occasion, although afterwards Sarah finds it very hard to watch the video of her mother.

One year later Sarah rings the counselor to say she is pregnant, she and Tom have decided to have a family after all.

useful model to use whenever the results of a test would give the client information about their future health, changing what is for them a potential situation into a certainty [14].

KEY PRACTICE POINT

A predictive test could be defined as any test that changes a potential health risk for the client into a certainty. This does not just apply to genetic tests, but could apply in many 'screening' situations, such as a renal ultrasound that detects cysts, or a full blood count that could detect sickle cell disease. Clients should be supported when undergoing such tests, and prepared for possible outcomes.

3.3 *Care of people affected by HD*

The needs of clients affected by HD are diverse, and require the input of professionals from many disciplines. A study [15] of affected individuals in a regional area showed that there were significant problems associated with communication between families and professionals in health and social care. Although some services were deficient, in many cases they existed but were accessed only after a crisis had occurred. A model of care that was introduced involves interdisciplinary working, but the identification of a named key professional is crucial to the success of such teamwork. The key worker takes responsibility for the communication of information from one professional to another, for the regular assessment of the clients' needs and for the organization of services at an early stage to reduce the necessity for crisis management, with all its inherent stresses on both the family and the services.

Harding Family

Peter Harding is 42 years of age. He has inherited HD from his father Cyril. Peter has been affected with HD for at least 6 years, although even before he was diagnosed he was very moody and prone to bouts of depression. These contributed to the divorce from his wife, before the reason was known.

Peter's wife Angie now offers him a lot of support, but she lives 2 miles from him, and has a job to support their two sons. For 2 or 3 years Peter met virtually no one from week to week. However, after he set fire accidentally to the kitchen curtains while having a fry-up late one night, social services became aware of his needs. He now attends a day center 3 days a week, where he does gardening and cooking and plays computer games. He also has speech therapy every week at the center to help him with his speech and swallowing difficulties.

4. Psychiatric conditions

4.1 Genetic influence on psychiatric conditions

The knowledge of genetic influences on psychiatric conditions has altered dramatically over the past 10 years. Because of the disabling effect of psychiatric illness on everyday living, the difficult social effects on the family, and the stigma that is still attached to such conditions, family members are often very concerned about the risk that such conditions 'may be passed on'.

Assessing the genetic influence on such conditions is difficult. These disorders are relatively common in the general population, with a lifetime risk of $\sim 1\%$ for either [16]. As they follow no clear Mendelian patterns of inheritance, they are clearly not due simply to an inherited gene mutation. In addition, it is possible that being raised by a parent with a psychiatric condition may influence the mental health of the offspring of that family, making it more likely that there are multiple cases within a family. However, studies of twins raised together or separately have shown that the children of parents with affective disorder and schizophrenia are more likely to develop these conditions than the offspring of unaffected parents.

4.2 Affective disorder (manic depression)

This is a term used to cover manic depression or depression alone. Studies indicate that the earlier the age at which a person is affected, the more likely it is that relatives will also be affected. However, the only basis we have upon which to advise a client inquiring about their own risk is the empirical data collected on families. These data enable a risk figure to be given, ranging from a risk of 5% if a second-degree relative has manic depression, to a risk of about 50% if both parents of the client have been affected.

4.3 Schizophrenia

Using data from empirical studies, it is clear that there is an increased risk of schizophrenia to relatives of an affected person. The genetic link is demonstrated when we consider the incidence of schizophrenia in twins. When one twin is affected, 40% of monozygotic twin siblings are also affected, as compared with only 10% of dizygotic twin siblings. This indicates that even after allowing for a common environment, the concordance between those who inherit identical genetic material risk is high.

Again, the risk is high (45%) if both parents of the client have schizophrenia.

Excellent tables for assessing risk are available in *Practical Genetic Counselling* by Peter Harper [16].

It is important when advising families of their risk to try to confirm the diagnosis in the affected person. Some inherited neurological conditions such as Huntington's disease or DRPLA may present with a severe psychiatric illness, but of course the risk to the client may be very different.

4.4 Presenile dementia (early Alzheimer's disease)

Presenile dementia is a condition (like breast or bowel cancer) that may occur sporadically, or may be caused by a gene mutation. Generally, late-onset Alzheimer's disease will be due to old age, rather than an inherited mutation, but if there is a history of dementia occurring in several younger members of a family, this may be connected with a mutation in one of the *presenilin* genes [17]. Where a gene mutation exists, children of an affected parent will be at 50% risk of inheriting the mutation and developing early dementia.

5. Genes and more common disease

The previous section of this chapter talked about conditions in which there is a strong link between a genetic mutation and the subsequent development of a disease. In the future, genetics may be more concerned with situations in which a genetic test shows that an individual has a genetic polymorphism that gives them a higher or lower chance of developing a particular disease, manifest the harmful effects of habit such as smoking, or that predicts the effectiveness of a specific pharmacological treatment. In this sort of situation, the positive and negative benefits of knowing that information need to be carefully weighed. This should include issues such as whether the test result gives any useful information about possible treatment, what the costs and benefit of the treatment are, is behavior change possible and what effects would it have, does the test result have implications for the rest of the family.

The examples used up until now have been concerned with relatively rare single gene disorders which in the main have been seen by specialized clinical genetics services. The field of cancer genetics, however, has led to greater collaboration between primary care, cancer services and genetics services in relation to guidelines for referral and screening for the early detection of cancer. In the same way that Huntington's disease was used as a model to consider factors important in designing predictive testing programs for adult-onset disorders, the development of cancer genetic services can act as a model for developing more integrated medical services when the concern is to identify at risk individuals and initiate screening and/or treatment.

6. Genetic hemochromatosis

It is useful to look in detail at a recessive genetic condition that is detectable and where effective treatment is possible. The opportunity to treat a condition changes the rationale for testing presymptomatically.

To help you think about some of the issues, we are going to work through the case of the Jones family.

CASE EXAMPLE   ANNE

Anne Jones rings up the genetics department in a great deal of distress. Her husband Mike is terminally ill with liver cancer. She is very angry because he had been complaining for some years of always being tired, pains in the joints in his fingers and generally feeling unwell. She said their life together had been very difficult over that time and one of her biggest regrets was that he had become impotent and this made her feel that their marriage was in danger of being over even though she still loved him. She and Mike had always felt that he was ill but no diagnosis had been made until he became jaundiced and was shown to have liver cancer. He was then told he had hemochromatosis, a genetic disorder of iron metabolism that meant he carried on absorbing iron from his food even when he had sufficient body iron stores. This iron overload had probably caused his symptoms and also had led to the development of his liver cancer. The reason Anne was so angry, however, was that she had been told that if he had been diagnosed with hemochromatosis before his liver damage had started his cancer could have been prevented by the simple treatment of having blood taken regularly. She had been onto the

---

Internet and read that hemochromatosis was a common genetic disease, it was recessive and one in ten people from Northern Europe carried the gene. She had worked out that if she carried the gene then her children were at a one in two chance of having the condition. She had to know if she carried the gene because she did not want her children to die of this terrible condition. She was also worried about Mike's sister and brother and had told them to go to their GP and be tested.

Think about how you would feel in this situation. What things would you want to know about before you had a test? If you were Mike's brother would you be anxious to be tested?

Some time later James, a cousin of Mike's is seen in the genetic center. As a result of extended family testing he has been shown to have two copies of the gene mutation that gives a risk of hemochromatosis and has been referred to discuss the risk to his children. He is fit and well, his GP has checked his ferritin and transferrin saturation and he shows no evidence of iron overload. He also is angry, and walks into the counseling session saying that he wished he never heard of the condition, he was absolutely fine but because he had the test for hemochromatosis he was having trouble getting life insurance. No one could tell him if he was going to run into problems with this disease or how many times he should see a doctor or how often he should have blood tests done. In fact the information he had been given was no use at all and he should never have had the test done.

Put yourself in James' position. What do you think about being tested for this condition now?

Genetic hemochromatosis is an autosomal recessive inherited disorder of iron metabolism. It is a treatable adult-onset disorder and screening is possible either using measures of serum iron, transferrin saturation is the most sensitive, or using a genetic test, although the value of using the genetic test for screening rather than family testing is not yet established.

The gene (*HFE*) was identified in 1996 [18] and two common mutations are shown to account for 90% of cases in the European population. Most affected people have two copies of a mutation called C282Y; a small proportion have the C282Y mutation together with a mutation called H63D. The genetic predisposition leads to accumulation of body iron stores over time.

Complications including cirrhosis, primary liver cancer, cardiomyopathy, arthritis and diabetes develop when iron overload is sufficient to cause organ damage [19]. Treatment for the condition involves removing multiple units of blood until the body iron stores return to normal. A maintenance program of venesection is then needed to prevent reaccumulation of iron. This treatment prolongs survival in symptomatic people and appears to restore normal life expectancy if started early enough in the course of the disease [20]. For this reason, arguments have been made proposing population screening as a way of diagnosing this disease early and preventing the complications of it.

If we consider the case example, someone in Anne's position might be very positive about universal screening for hemochromatosis. If her husband had been diagnosed early enough then his cancer would have been prevented and he would not be dying. However, population screening, as discussed in *Chapter 1*, requires a stringent analysis of potential costs and benefits because it involves taking a population who have no medical complaints and identifying some who may become ill. The main problem with screening for hemochromatosis is that the natural history of the condition is not known. It is clear that individuals identified both by screening for abnormal iron metabolism (transferrin saturation) or by using a DNA based test do not inevitably develop complications from having the predisposition. The true risk of developing serious complications is not known. In addition, it is not clear what strategy should be used for screening, biochemical or genetic, or when and how screening programs should be carried out. There is also legitimate concern about possible 'genetic discrimination', that is, effects on insurance, employment, etc. if an individual is identified as having the predisposition for developing complications of this condition. For this reason population screening is not recommended at the present time. However, genetic hemochromatosis could prove to be an important situation in which to consider the possible positive and negative effects of using genetic testing to identify disease predisposition for the purpose of treatment or prevention.

7. Conclusion

In this chapter, we have described the influence of genes on three very different types of adult-onset disease. Genetic hemochromatosis is treatable if detected in the early stages, whereas screening and early treatment can influence the outcome for those at high risk of familial cancer. At present there is no cure for HD and other neurodegenerative disorders.

Psychological care of adult patients is as important as genetic information, as guilt, blame, anxiety and hopelessness may accompany the feeling of risk in any of these situations. As expressed earlier, it is not necessarily the risk but the perceived burden of the disease that is important to the family. Adults with first-hand experience of a disease will perceive it differently from those who have only been told about the condition by others. The use of counseling skills in any genetic situation helps the practitioner to understand the client's perspective and therefore to offer appropriate support.

TEST YOURSELF

Q1. A woman attending her GP surgery is aware of the risk of bowel cancer. She refuses colonoscopy, and refuses to allow her siblings (who are all at risk) to be informed. How could the practice nurse approach this situation?

Q2. Is it worthwhile undertaking genetic tests for a man whose father had genetic hemochromatosis? Explain your response.

Q3. Prenatal testing for a *BRCA1* mutation (a gene fault that predisposes a woman to breast and ovarian cancer) is technically possible in some families with a strong history of breast cancer. In your opinion, is termination of pregnancy for such a condition warranted?

References

1. Bertwistle D, Ashworth A (1999) The pathology of familial breast cancer: How do the functions of BRCA1 and BRCA2 relate to breast tumour pathology? *Breast Cancer Res* **1** (1): 41–47.
2. Knudson A-GJ (1971) Mutation and cancer: statistical study of retinoblastoma. *Proc Natl Acad Sci USA* **68** (4): 820–823.
3. Claus EB, Risch N, Thompson WD (1994) Autosomal dominant inheritance of early-onset breast cancer. Implications for risk prediction. *Cancer* **73** (3): 643–651.
4. Eccles DM, Evans DG, Mackay J (2000) Guidelines for a genetic risk based approach to advising women with a family history of breast cancer. UK Cancer Family Study Group (UKCFSG). *J Med Genet* **37** (3): 203–209.
5. Hendrick RE, Smith RA, Rutledge JH, Smart CR (1997) Benefit of screening mammography in women aged 40–49: a new meta-analysis of randomized controlled trials. *J Natl Cancer Inst Monogr* **22**: 87–92.
6. Tinley ST, Lynch HT (1999) Integration of family history and medical management of patients with hereditary cancers. *Cancer* **86** (11 Suppl.): 2525–2532.
7. Lovett E (1976) Family studies in cancer of the colon and rectum. *Br J Surg* **63** (1): 13–18.
8. Lynch J, Lynch HT (1994) Genetic counseling and HNPCC. *Anticancer Res* **14** (4B): 1651–1656.
9. Horowitz M, Wilner N, Alvarez W (1979) Impact of Event Scale: a measure of subjective stress. *Psychosom Med* **41** (3): 209–218.
10. Lanska DJ (2000) George Huntington (1850–1916) and hereditary chorea. *J Hist Neurosci* **9** (1): 76–89.
11. Gusella JF, MacDonald ME, Ambrose CM, Duyao MP (1993) Molecular genetics of Huntington's disease. *Arch Neurol* **50** (11): 1157–1163.
12. Craufurd D, Tyler A (1992) Predictive testing for Huntington's disease: protocol of the UK Huntington's Prediction Consortium. *J Med Genet* **29** (12): 915–918.
13. International Huntington's Disease Association (2002) Guidelines for the molecular genetic predictive test in HD. http://www.huntington-assoc.com/guidel.htm

14. Harper PS, Lim C, Craufurd D (2000) Ten years of presymptomatic testing for Huntington's disease: the experience of the UK Huntington's Disease Prediction Consortium. *J Med Genet* **37** (8): 567–571.

15. Skirton H, Glendinning N (1997) Using research to develop care for patients with Huntington's disease. *Br J Nurs* **6** (2): 83–90.

16. Harper PS (1998) *Practical Genetic Counselling*, 5th edn. Oxford: Butterworth-Heinemann.

17. Poorkaj P, Sharma V, Anderson L *et al.* (1998) Missense mutations in the chromosome 14 familial Alzheimer's disease *presenilin 1* gene. *Hum Mutat* **11** (3): 216–221.

18. Feder JN, Gnirke A, Thomas W *et al.* (1996) A novel MHC class I-like gene is mutated in patients with hereditary haemochromatosis. *Nat Genet* **13** (4): 399–408.

19. Niederau C, Fischer R, Purschel A, Stremmel W, Haussinger D, Strohmeyer G (1996) Long-term survival in patients with hereditary hemochromatosis [see comments]. *Gastroenterology* **110** (4): 1107–1119.

20. Niederau C, Strohmeyer G, Stremmel W (1994) Epidemiology, clinical spectrum and prognosis of hemochromatosis. *Adv Exp Med Biol* **356**: 293–302.

Further reading

Bennett RL (1999) *The Practical Guide to the Genetic Family History*. New York: Wiley Liss. (Detiled information, especially on cancer syndromes.)

Eeles R, Ponder BA, Easton DF, Horwich A, McVie G (editors) (1996) *Genetic Predisposition to Cancer*. London: Chapman and Hall. (Full coverage of genetic influences on cancer.)

Harper PS (1998) *Practical Genetic Counselling*, 5th edn. Oxford: Butterworth-Heinemann. (Adult-onset conditions covered, including predictive testing.)

10 Development of genetic services and the genetic counseling profession

1. Introduction

Genetic services have developed as a specialist service in health care over the past 50 years. Clinical genetic services were established in both Europe and in North America in the post-war period, but substantial growth occurred in the 1980s, as recombinant DNA technology opened the door to a number of new tests and services for families. In many countries, including the UK, Belgium, Holland, Australia Canada and the USA, the service is provided by medical geneticists and nonmedical genetic counselors, working together in teams. The specific remit of genetic services is to provide information and support for individuals and families at risk of or affected by a genetic condition.

2. A matter of peas

It is difficult to identify with any accuracy the birth of the profession of genetic counseling in Europe, but most people would acknowledge Gregor Mendel as the father of genetics. He spent years patiently breeding peas, counting the offspring, using the hybrids to produce more plants, and counting the resultant plants. His meticulous work led him to discover and describe accurately the mode of recessive inheritance. However, although Mendel laid the basis for genetic theory, it was Francis Galton who began to test the theories in relation to humans. Mendel and Galton were born the same year (1822). Mendel's scientific pursuits were somewhat restricted by his theological vocation, and perhaps by the lack of acknowledgement he received in his lifetime, whereas Galton became one of the Victorian polymaths. Galton was the cousin of Charles Darwin, and the publication in 1859 of *The Origin of Species* [1] had influenced his thoughts on matters of heredity and his religious beliefs. In the retrospective light of current scientific method, Galton's studies were primitive and naïve, however, perhaps the greatest value of Galton's work lies in his interest in the new territory of human genetics. His varied studies included work on the hereditary nature of ability [2,3] and the numerical assessment of risk [4]. Galton's interest in inherited traits continued to influence his scientific ventures, and eventually he realized that the traits were somehow inherited from both parents, and that they were transferred in the fertilized ovum. He experimented using sweetpeas independently of Mendel, and performed twin studies.

The outcome of Galton's research and thinking on these matters was the development of the theory of eugenics. He proposed that as characteristics necessary for the improvement of the species are largely inherited, it made sense to encourage breeding of the 'fitter' members of a society, and to discourage those 'less fit' from having children. In this way any society could be enhanced, for although the struggle for survival applies equally to all species, the knowledge of inheritance confers upon man the opportunity to influence his own evolution. Galton regarded eugenics as a science in its own right, a theory which was morally superior to natural selection. He felt natural selection relied upon over-production and subsequent destruction of excessive stock. By enhancing conditions for reproduction of healthy stock, Galton felt not only society, but the individuals within that society, would benefit [5]. This was a theory which had its own morality, based as it was on the belief that all those born into society ought to be able to make a positive contribution, thus resulting in a better society in the long-term. However, it relied upon the subjective judgments of certain individuals (self-selected) of the worth of other individuals.

3. The eugenics movement

The Eugenics Society was born of Galton's ideas. Although Galton emphasized the need for such judgment of human fitness to be based on statistical evidence, this type of evidence is unobtainable by ethical means. As Blacker [6], a prominent member of the society later said, as both genetics and environment have an influence on the eventual ability of a person to contribute positively to society, the judgment can only be made where environmental influence is absolutely equal, therefore making such a judgment is impossible. Blacker also stressed that the Eugenics Society was occupied with the study rather than the practice of eugenics.

The eugenics movement certainly had its critics, one of the most vocal being GK Chesterton, who felt that the principles of eugenics increased the burden of the working class by inhibiting reproductive choice. At the time, it was not only those with genetic conditions who were considered 'unfit' to breed. Many acquired medical conditions such as syphilis, tuberculosis and alcoholism were thought to be inherited, and poverty, criminality and prostitution were also indicators of unfitness.

A Bill concerning voluntary sterilization for those in the Social Problem Group was drafted by the Brock Committee in 1934, but did not become legislation. One of the aims of the Eugenics Society [6] was to make affordable contraceptive methods available to those who wished to have them, but the Brock Committee legislation differed in that it was based on a judgment of fitness rather than personal choice.

The abhorrence of eugenic principles felt by some individuals was very probably exacerbated by the distortion and abuse of the original ideals by those with completely different motives from those of the founders of the Eugenics Society. The notoriety of eugenics reached a zenith during the period of the Third Reich in Germany in the 1930s. Hitler had proclaimed his belief in the supremacy of the Aryan peoples in *Mein Kampf*. His suggestion that the State had both a moral and practical responsibility to ensure that only those who were genetically desirable should procreate culminated in plans to sterilize sufferers and carriers of hereditary diseases. In 1933, a compulsory sterilization law was passed, which included those with hereditary deafness, blindness and Huntington's disease among other congenital disabilities. Later in the same decade, these disabilities became grounds for extermination under the same regime. However, compulsory sterilization for genetic disease has not been solely confined to Germany, it has also been practiced in the United States and in a number of Scandinavian countries [7].

4. Growth of medical genetics

The first episode of genetic counseling is said by Lynch and Lynch [8] to have taken place in 1895, when a doctor discussed the risk of familial cancer with a family seamstress who had a strong family history of the disease. The doctor reported that the seamstress looked sad, and when questioned she confessed that she felt her time on earth was limited, due to the frequent occurrence of cancer in her family. She proved to be right, dying young of the condition.

However, the importance of genetics to medicine only became generally appreciated with the discovery of the genetic basis of disease early in the 20th century. One of the first publications on what would now be termed genetic counseling appeared in 1933, titled *The Chances of Morbid Inheritance*. Blacker wrote the book to assist doctors in advising patients about the chances of having a child with an inherited condition. The emphasis was placed on informing couples on the *advisability* of having children. In view of the inability of medical science to detect problems prenatally, or to treat the majority of conditions, the options for couples at risk were limited. An extension of the Eugenic Society's work in this field occurred just prior to the Second World War, when it undertook to offer advice to people who were seeking information on their genetic risks, prior to marriage or having a family. However, the suspicion with which modern genetic counseling is viewed by some couples may have stemmed from this approach.

One doctor who recognized the importance of genetics in disease causation was Archibald Garrod [9] who built upon Mendel's discoveries to determine the autosomal recessive inheritance of a number of conditions termed 'Inborn Errors of Metabolism'. Garrod was able to correctly predict the 25% risk of inheriting the disease for children of two carrier parents. He also comprehended the relevance of consanguineous marriage to recessive inheritance of disease.

In 1931, Lancelot Hogben, a Professor of Social Science, proposed the establishment of dedicated units for the study of human genetics, staffed by teams comprising geneticists, statisticians, doctors, psychologists and ethnologists. Hogben had grasped the complexity of the subject of clinical genetics and its probable influence on individual human existence. His call for qualified workers in a number of disciplines is the model for clinical genetic services today.

5. Establishment of clinical genetic services

The first clinical department of genetics in the UK was established at The Hospital for Sick Children, Great Ormond Street, London, and a geneticist was employed for clinical duties within the hospital from 1951. At the time it was believed that genetic services were needed to help parents to decide whether to have further children, and to alert medical staff to risks to the subsequent offspring of those parents, so that early diagnosis and treatment were possible [10]. Genetic services were also established in the USA in the same decade, and it was a US geneticist, Sheldon Reed, who first used the phrase 'genetic counseling' in 1947 [11]. Clinical genetic services are now a regular part of healthcare provision in the First World. Accurate human chromosomal studies have been possible since the late 1950s; these have enabled syndromes caused by chromosomal imbalance to be accurately identified. Further stimulus for expansion arrived with the advent of recombinant DNA technology in the 1970s. Although clinical examination of patients is still important, gene mutation analysis is increasingly helpful in clarifying the genetic diagnosis, and making prenatal and presymptomatic and predictive testing available to families. Increasing awareness of the psychological impact of genetic conditions and risk upon families has led to the inclusion of psychological support as an integral part of genetic services.

Genetic services are cited in the first report to the NHS Central Research and Development Committee on the new genetics [12] as being of continuing importance as centers of expertise, even if some genetic counseling is performed in the future in primary care or other departments.

6. Genetic services – the current situation

The scientific progress in genetics has led to 'a new taxonomy of disease' [13], meaning that the discovery of the underlying genetic cause of a disease can be used as a basis for understanding the effects of the disease and for developing treatment strategies. This statement signals a startling change from the traditional approach, whereby symptoms and effects of the disease were used as the starting point for development of treatment. The current definition of clinical genetic services [14] includes diagnosis of the condition, the prevention or amelioration of the disease, information about risk and reproductive choice. Although technology has enhanced the diagnostic skill, risk assessment, treatment and options, it appears that the underlying human rationale for the service has not been clouded by pure technology. However, it is clear that the forces influencing the development of clinical genetic services have derived first from an academic, and then from a medical model. The model of service has now been further modified because of the influence of genetic counselors, who see a primary part of their role as providing support to families.

Although clinical genetics is established across Europe and North America, the quality and type of service provision varies enormously. Most teams are led by medical geneticists, in some areas medical personnel are the sole providers. In many European countries (e.g. Poland, Italy and France) genetic services are provided almost exclusively by medically trained staff, with very little input from other professions, such as nursing. However, in a handful of countries other professionals are employed within the team. These are known as genetic counselors, and many have a background in nursing or similar paramedical field. Whereas in the early days of the specialty, genetic counselors trained 'on the job', increasingly they prepare via a Master's degree in genetic counseling.

7. Establishment of a genetic counseling profession

Is genetic counseling a profession? The main attributes of a profession have been described as (i) the use of skills based on specialized knowledge, (ii) entry to the profession via an examination to demonstrate competence, (iii) adherence to a code of conduct, (iv) establishment of a professional organization, and (v) practice that benefits the common good. The core competencies for the practice of genetic counselling are outlined in *Figure 1*. Clearly, genetic counselors require specialised knowledge to practice, and genetic counseling benefits the common good, although it does this by attending to the specific needs of individuals. In the USA, a clear professional pathway has been established for many years, with a Board accreditation system for practitioners. However, in most countries in Europe there are no specific codes of conduct, professional bodies or entry criteria in place. This has prompted moves to establish conditions for genetic counseling to grow as a profession in the UK.

In the UK, teams of medically trained clinical geneticists and nonmedically trained genetic counselors in NHS regional centers provide genetic services. The philosophy of these teams is based on the belief that individuals have a right to be properly informed about the genetic risks and reproductive options, and that they should be supported during any decision-making process [15]. Genetic counselors have traditionally provided support for clients but increasingly they are accountable for their own caseload and work autonomously within the team as well as supporting the work of medical team members.

The genetic counselor is able to:

Communication skills

1. Establish a relationship with the client and elicit the client's concerns and expectations.
2. Elicit and interpret appropriate medical, family and psychosocial history.
3. Convey clinical and genetic information appropriate to the client's individual clinical needs.
4. Explain the options available to the client, including risks, benefits and limitations of such options.
5. Document information including case notes and correspondence in an appropriate manner.
6. Plan, organize and deliver professional and public education.

Interpersonal, counseling and psychosocial skills

7. Identify and respond to emerging issues of a client or family.
8. Acknowledge the implications of individual and family experiences, beliefs, values and culture for the genetic counseling process.
9. Make a psychosocial assessment of client's needs and resources and provide support, ensuring referral to other agencies as appropriate.
10. Use a range of counseling skills to facilitate client's adjustment and decision-making.
11. Establish effective working relationships to function within a multidisciplinary genetics team and as part of the wider health and social care network.

Ethical practice

12. Recognize and maintain professional boundaries.
13. Demonstrate reflective skills within the counseling context and in personal awareness for the safety of clients and families by participation in counseling/clinical supervision.
14. Practice in accordance with the AGNC Code of Ethical Conduct.
15. Present opportunities for clients to participate in research projects in a manner that facilitates informed choice.
16. Recognize his or her own limitations in knowledge and capabilities and discuss with colleagues or refer clients when necessary.
17. Demonstrate continuing professional development as an individual practitioner and for the development of the profession.
18. Contribute to the development and organization of genetic services.

Critical thinking skills

19. Make appropriate and accurate genetic risk assessment.
20. Identify, synthesize, organize and summarize relevant medical and genetic information for use in genetic counseling.
21. Demonstrate the ability to organize and prioritize a caseload.
22. Identify and support clients' access to local, regional and national resources and services.
23. Develop the necessary skills to critically analyze research findings to inform practice development.

Figure 1. Core competencies for the practice of genetic counseling.

The four main components of the role of genetic counselor are:

- Communication with clients, both to obtain the family medical history necessary to provide the client with reliable information, convey the genetic information, and present the options available to the family in a nonjudgmental manner.
- Client support, particularly at times of decision-making or particular stress, e.g. after a new diagnosis has been made in the family, during an at-risk pregnancy or when testing is being considered.
- Education of clients and other health professionals on issues related to clinical genetics.
- Skilled interpretation of current research findings for the benefit of clients.

The Association of Genetic Nurses and Counselors in the UK has established a formal registration process for genetic counselors. A code of ethics has been adopted (*Figure 2*), and the registration system is based upon a set of competency standards, which are the knowledge, skills and attitudes necessary to enable a practitioner to function effectively and safely in a specific field of activity. Standards of competence facilitate the regulation of practitioners and provide a useful basis for devising educational pathways that are directly applicable to the skills required for the work.

It is apparent that the demand for genetic services will increase as technology offers more in the form of genetic testing, pharmacogenetics and possibly gene therapy. In addition to the rare genetic disorders, genetic services may also be required to make a contribution to the health care of those with common conditions in which there is a genetic component. This may not be direct care for patients with those conditions, but will almost certainly involve providing education and consultation for other health care professionals [16].

The genetic counselor needs to be educated sufficiently to interpret the complex scientific data for the benefit of patients, and to possess the counseling skills to ensure clients can express their opinions, explore the options and make informed decisions. Courses that address the specific training needs of genetic counselors are still scarce. As genetic counseling is overwhelmingly a practical clinical exercise, training programs must include opportunities for observation of good practice and supervised clinical practice for the student.

As consumers become more aware of the availability of genetic counseling and testing, and clinical governance becomes more rigorous, genetic counselors must be prepared to provide a certain standard of service.

9. Conclusion

The new profession of genetic counseling is becoming established to provide care, support and information for families at risk of a genetic condition. However, increasingly genetics is a core component of everyday health care,

(a) Self-awareness and development

Genetic counselors should:

- Recognize the limits of their own knowledge and abilities in any given situation, and decline any duties or responsibilities that cannot be carried out in a safe and competent manner;
- Be responsible for their own physical and emotional health as it impacts on their professional performance;
- Report to an appropriate person or authority any conscientious objection that may be relevant to their professional practice;
- Maintain and improve their own professional education and competence.

(b) Relationships with clients

Genetic counselors should:

- Enable clients to make informed independent decisions, free from coercion;
- Respect the client's personal beliefs and their right to make their own decisions;
- Respect clients, irrespective of their ethnic origin, sexual orientation, religious beliefs and gender;
- Avoid any abuse of their professional relationship with clients;
- Protect all confidential information concerning clients that is obtained in the course of professional practice: disclosures of such information should only be made with the client's consent, unless disclosure can be justified because of a significant risk to others;
- Report to an appropriate person or authority any circumstance, action or individual that may jeopardize the care, health or safety of the client;
- Seek all relevant information required for any given client situation;
- Refer clients to other competent professionals if they have needs outside the professional expertise of the genetic counselor.

(c) Relationships with colleagues

Genetic counselors should:

- Collaborate and cooperate with other colleagues in order to provide the highest quality of service to the client;
- Foster relationships with other members of the clinical genetics team, to ensure that clients benefit from a multidisciplinary approach to care;
- Assist colleagues to develop their knowledge of clinical genetics and genetic counseling;
- Report to an appropriate person or authority any circumstance or action which may jeopardize the health and safety of a colleague.

(d) Responsibilities within the wider society

Genetic counselors should:

- Provide reliable and expert information to the general public;
- Adhere to the laws and regulations of society. However, when such laws are in conflict with the principles of the profession, genetic counselors should work toward change that will benefit the public interest;
- Seek to influence policy makers on human genetic issues, both as an individual and through membership of professional bodies.

Figure 2. AGNC Code of Ethical Practice.

and professionals from many disciplines and specialties will need to be educated about genetics to work competently and confidently with clients. Skill in counseling, and an awareness of the potential ethical, psychological and social issues are also a necessity for competent practice.

References

1. Darwin C (1998) *The Origin of Species*. New York: Gramercy.
2. Galton F (1874) *English Men of Science, their Nature and Nurture*. London: Macmillan.
3. Galton F (1892) *Hereditary Genius*, 2nd edn. London: Macmillan.
4. Galton F (1987) The average contribution of each several ancestor to the total heritage of offspring. *Royal Society Proceedings*.
5. Galton F (1908) *Memories of My Life*. London: Methuen & Co.
6. Blacker C (1945) *Eugenics in Retrospect and Prospect*. Glasgow: University Press.
7. Kevles DJ (1995) *In the Name of Eugenics*, 2nd edn. New York: Alfred A Knopf/Harvard University Press.
8. Lynch J, Lynch HT (1994) Genetic counseling and HNPCC. *Anticancer Res* **14** (4B): 1651–1656.
9. Garrod A (1909) *Inborn Errors of Metabolism*. London: Hodder and Stoughton.
10. Carter C, Fraser Roberts J, Evans K, Buck A (1971) Genetic clinic: a follow-up. *Lancet* **February 6**: 281–285.
11. Reed S (1955) *Counselling in Medical Genetics*. Philadelphia: W.B. Saunders.
12. Department of Health (1995) *Report of the Genetics Research Advisory Group. A First Report to the NHS Central Research and Development Committee on the New Genetics*. London: Department of Health.
13. Department of Health (1995) The Genetics of Common Diseases. A Second Report to the NHS Central Research and Development Committee on the New Genetics. London: Department of Health.
14. Ad Hoc Committee on Genetic Counselling. American Society for Human Genetics (1975) Genetic counselling. *Am J Hum Genet* **27**: 240–242.
15. Clarke A (1990) Genetics, ethics, and audit. *Lancet* **335** (8698): 1145–1147.
16. Skirton H, Barnes C, Curtis G, Walford-Moore J (1997) The role and practice of the genetic nurse: report of the AGNC Working Party. *J Med Genet* **34** (2): 141–147.

Further reading

Kevles DJ (1995) *In the Name of Eugenics*, 2nd edn. New York: Alfred A Knopf/Harvard University Press.

Marteau T, Richards M (editors) (1996) *The Troubled Helix: Social and Psychological Aspects of the New Human Genetics*. Cambridge: Cambridge University Press.

Restall R. Genetic counseling: coping with the human impact of genetic disease.
http://www.accessexcellence.org/AE/AEC/CC/counseling_background.html

Appendix I – Website resources

Increasingly, the source of information used by both professionals and families is the worldwide web. The following are sites that may be helpful to you. As sites may change rapidly we are unable to verify their reliability, so ensure you check them out, especially if recommending one to a family in your care.

Website	Description	Use
Professional organizations		
http://www.bshg.org	British Society for Human Genetics Organization for all health professionals and scientists involved in genetics – whether clinical practice or research List of genetic centers across UK	Professional
http://www.agnc.org.uk	Association of Genetic Nurses and Counsellors UK-based professional organization Information on training and registration system for genetic nurses and counselors, code of ethics, standards of competence	Professional
http://nursing.creighton.edu/isong/practice/credential.htm	International Society of Nurses in Genetics (ISONG) For nurses in genetics health care, largely US oriented	Professional
http://www.nsgc.org	National Society of Genetic Counselors website Information on genetic counselors in the USA	Professional
http://www.faseb.org/genetics/ashg/ashgmenu.htm	American Society of Human Genetics Primary professional membership organization for human geneticists in North America. Members include researchers, academics, clinicians, laboratory practice professionals, genetic counselors, nurses and others involved in human genetics	Professional
http://www.hgsa.com.au/asgc/about.html	Australasian Society of Genetic Counsellors	Professional
http://www.bac.co.uk/	British Association for Counselling and Psychotherapy Valuable information on practice in counseling, including standards of practice and code of ethics	Professional
http://www.counseling.org/	American Counseling Association Useful information on aspects of counseling practice, including ethical practice	Professional

General information

URL	Description	Category
http://www.medinfo.cam.ac.uk/phgu/default.asp	Public Health Genetics Unit. Useful resource for genetics and public health with links to other websites and reviews of specific areas, such as genetics and insurance	Professional
http://helios.bto.ed.ac.uk/bto/glossary/	Comprehensive glossary of genetic terms	Professional
http://www.doh.gov.uk/nsc/	National screening committee. Government-sponsored website relating to screening policy within the UK	Professional
http://www.cdc.gov/genetics/	Center for disease control in Atlanta. American website focusing on public health and genetics	Professional
http://www.yourgenome.org/	General website on genetic issues sponsored by the human genome project	Professional/family
http://www.accessexcellence.org/AE/AEC/CC/counseling_background.html	Article by Robert Resta on history of genetic counseling	Professional
http://www.bshg.org.uk/Society/CGS/CGSPedSheets.pdf	Clinical Genetics Society site on pedigree drawing	Professional
http://www.fccc.edu/research/programs/advisors/knudson/	Knudson's two-hit theory explained	Professional
www.nhs.uk/nelh	NHS national electronic library for health. Access to clinical evidence, guidelines, press and website reviews	Professional
http://www.ich.ucl.ac.uk/cmgs/bayes99.htm	Institute of Child Health site for training in molecular genetics. Explanation and practice examples of Bayes theorem	Professional
http://www.nas.com/downsyn/parent.html	Parent matching and support groups. Parental and sibling support. Addresses for several conditions	Family
http://www.nlm.nih.gov/	National Library of Medicine. For research and Medline	Professional
http://www.cancergenetics.org/genetics.htm	Cancer and genetics. General information on cancer	Family

Website	Description	Use
http://audumla.mdacc.tmc.edu/~oncolog/09_Familial.html	Familial cancer syndromes and research General information site	Professional
General sites for disease information		
http://www.nfdht.nl/guidelines.htm	International Cancer Guidelines Guidelines for management of HNPCC screening	Professional
http://www.aap.org/	American Academy of Pediatrics Search engine, plus research information	Professional
http://www.autism.org/	Center for the Study of Autism Information on all syndromes with features of autism	Family/professional
http://www.kumc.edu/gec/support/	Genetic and rare conditions site Information on many disorders	Professional
http://info.ki.se/index_en.html	Karolinska Institutet, Sweden Search engine for latest research on a large number of rare and common conditions	Professional
http://www.mdchoice.com/pt/index.asp	Medical search engine	Family/professional
http://www3.ncbi.nlm.nih.gov/Omim/	Online Mendelian inheritance in man Online version of OMIM a catalog of inherited disease. Resource intended for specialist medical and health professional use	Professional
http://www.diseases.nu/index.htm	National Disease Information Center Information on physical diseases and mental disorders	Professional
http://www.nih.gov/	National Institute of Health A huge resource for genetic conditions and scientific research	Professional
http://www.rarediseases.org/	National Organization for Rare Disorders Database	Professional

Adult polycystic kidney disease		
http://www.kidney.ca/poly-e.htm	Polycystic kidney disease (PKD) Very clear and concise information	Family
http://www.mdchoice.com./pt/ptinfo/pkid.asp	Polycystic kidney disease User-friendly information	Family
http://www.pkdcure.org/	Polycystic Kidney Research Foundation Lots of general information, links, support groups, news, research	Family/professional
http://www.mc.vanderbilt.edu/peds/pidl/rephro/polykidn.htm	Vanderbilt Medical Center Lots of technical information	Professional
Angelman's syndrome		
http://asclepius.com/angel/	Angelman's syndrome Information for families and professionals	Family/professional
http://shell.idt.net/~julhyman/angel.htm	Angels Among Us What it's like to have an 'angel' in your family. Addresses, information, links Family/professional	Family
http://www.asclepius.com/angel/asfinfo.html	Facts about Angelman's syndrome Information, links, references	Family/professional
Chromosome 22q		
http://www.genes.uchicago.edu/telomere/22q.html	Chromosome 22q telomere References on this subject	Professional
http://www.sanger.ac.uk/cgi-bin/c22_diseases_table.pl	Chromosome 22 disease list List of all diseases known to map to chromosome 22, with OMIM numbers	Professional

Website	Description	Use
http://www.nt.net/~a815/chr22.htm	Chromosome 22 central Parent support group for chromosome 22 related disorders Information, links, support	Family
http://www.ucfs.net	UK 22q11 Support group site	Family
Cleft lip and palate		
http://www.clapa.mcmail.com/	Cleft Lip and Palate Association Comprehensive guide to information, medical care, links	Family/professional
http://www.cleft.org/	Smiles. A cleft lip and palate support group Support group, with friendly information	Family/professional
http://www.widesmiles.org/Default.htm	Wide smiles Cleft lip and palate resource Comprehensive guide to information, medical care, links	Family/professional
Cystic fibrosis		
http://www.cff.org/	Cystic Fibrosis Foundation Latest news, research, information	Family/professional
http://vmsb.csd.mu.edu/~5418lukasr/cystic.html#C	Cystic fibrosis index of on-line resources Lots of links to other pages connected with cystic fibrosis	Family/professional
http://cysticfibrosis.com/	Cystic fibrosis Medical information, plus good children's page	Family/professional
http://www.cysticfibrosis.co.uk/game.htm	The Creon game! Game to download for children to encourage them to take their enzyme capsules with food	Family

Down syndrome		
http://www.nads.org/	National Association for Down Syndrome (NADS) Counseling and support service for parents of children with Down syndrome	Family
http://www.ndss.org/	National Down Syndrome Society A comprehensive, on-line information source about Down syndrome	Family
http://www.ds-health.com/ds_sites.htm	Recommended Down syndrome sites on the Internet Quick access to lots of sites	Family/professional
http://www.nas.com/downsyn/welcome.html	Welcoming babies with Down syndrome Friendly site for families who have a child with Down syndrome	Family use
Familial adenomatous polyposis (FAP)		
http://www.mtsinai.on.ca/familialgican/FAPEnglish/fap.html	FAP. A guide for families Information on all aspects of FAP	Family
http://www.geneclinics.org/profiles/fap/	Gene clinics: FAP Lots of factual information, fairly readable	Family/professional
http://intouch.cancernetwork.com/journals/oncology/o9601e.htm	Genetic testing and counseling in familial adenomatous polyposis On-line paper, concise information	Professional
http://www.ncgr.org/gpi/odyssey/colon/fap_arts.html	Scientific articles: familial adenomatous polyposis Useful articles	Professional
Familial breast/ovarian cancer		
http://norp5424b.hsc.usc.edu/fbcpubs.html	Familial breast cancer publications Links to recent publications	Professional
http://www.familycancer.org/FamilyCancer/index_gene.stm	Family cancer, and genetic testing Information, links and support groups for breast and ovarian cancer	Family

Website	Description	Use
http://rpci.med.buffalo.edu/departments/gynonc/grwp.html	Gilda Radner Familial Ovarian Cancer Registry Comprehensive information	Family
http://www.gyncancer.com/ovarian-cancer.html	Ovarian cancer Technical and basic information	Professional
Hemochromatosis		
http://www.ghsoc.org/home.html	UK Haemochromatosis Society home page	Professional/family
Hereditary motor sensory neuropathies (HMSN) (also called Charcot Marie Tooth disease)		
http://www.cmt.org.uk/	Charcot Marie Tooth International UK Support group, with basic information	Family
http://www.ultranet.com/~smith/CMTnet.html	CMTnet Information on research and treatment	Family/professional
http://www.neuro.wustl.edu/neuromuscular/time/hmsn.html	Hereditary motor sensory neuropathies, Charcot Marie Tooth disease Range of disorders covered, technical information	Professional
http://www.muscular-dystrophy.org.uk/information/Key%20facts/hmsn.html	Muscular dystrophy campaign User-friendly information on HMSN	Family
Hereditary nonpolyposis colorectal cancer (HNPCC)		
http://www.cancergenetics.org/hnpcc.htm	Hereditary nonpolyposis colon cancer Questions and answers on a case study	Family
http://www.mdacc.tmc.edu/~hcc/	MD Anderson Cancer Center Hereditary colon cancer Basic information	Family/professional

186

URL	Description	
http://www.penrosecancercenter.org/info-hnpcc.htm	Penrose Cancer Center/hereditary nonpolyposis colon cancer (HNPCC) questions and answers User-friendly language	Family
http://www.ncgr.org/gpi/odyssey/colon/hnpcc_at.html	Scientific articles: hereditary nonpolyposis colorectal cancer Useful articles	Professional
Huntington's disease		
http://www.kumc.edu/hospital/huntingtons/	Caring for people with Huntington's disease Information, links	Family/professional
http://www.neuro-chief-e.mgh.harvard.edu/mcenemy/facinghd.html	Facing Huntington's disease Useful information	Family
http://www.hdac.org/	Huntington's Disease Advocacy Center Articles, stories, chatroom, links	Family
http://www.hda.org.uk/	UK's Huntington's Disease Association Information, news, events	Family/professional
Hypertrophic cardiomyopathy		
http://www.hcma-heart.com/	Hypertrophic Cardiomyopathy Association	Family/professional
http://www.kanter.com/hcm/	Hypertrophic Cardiomyopathy Association Symptoms, treatment, information	Family/professional
http://www.gilead.org.il/hcm/	The management of hypertrophic cardiomyopathy (review article) Online article	Professional
Neural tube defects		
http://hna.ffh.vic.gov.au/phd/folate/hlthprof.htm	Folate, a guide for health professionals Information biased towards Australian population	Professional

Website	Description	Use
http://www.eatwellmd.org/folic.htm	Maryland Dietetic Association Fact sheet on folic acid	Family
http://thearc.org/faqs/folicqa.html	Prevention of neural tube defects Comprehensive information on neural tube defects	Family
http://cpmcnet.columbia.edu/texts/gcps/gcps0052.html	Screening for neural tube defects – including folic acid/folate prophylaxis Up-to-date information	Professional
Neurofibromatosis		
http://www.aap.org/policy/00923.html	American Academy of Pediatrics committee on genetics health supervision for children with neurofibromatosis	Professional
http://www.neurosurgery.mgh.harvard.edu/NFR/	Neurofibromatosis resources. Site set up by person with NF2, information on neurofibromatosis, plus links.	Family
http://www.nfinc.org/	Neurofibromatosis, Inc. Information, support, links	Family
http://www.nf.org/	The National Neurofibromatosis Foundation Inc. Information, links, support groups	Family/professional
Prader–Willi syndrome		
http://www.pwsausa.org/postion/HCGuide/HCG.htm	Healthcare guidelines for individuals with Prader–Willi syndrome Recommendations and references	Professional
http://www.icondata.com/health/pedbase/files/PRADER-W.HTM	Prader–Willi syndrome Full run down of clinical features, plus history of disorder	Professional
http://www.geneclinics.org/profiles/pws/	Prader–Willi syndrome People-friendly information on disorder	Family/professional

URL	Description	Category
http://www.pwsausa.org/	The Prader–Willi Syndrome Association (USA) Information, links, support groups	Family
Smith–Lemli–Opitz syndrome		
http://www.med.jhu.edu/CMSL/SLOS.html	Smith–Lemli–Opitz syndrome Factual information	Professional
http://www.geneclinics.org/profiles/slo/	Smith–Lemli–Opitz syndrome User-friendly information on syndrome	Family/professional
http://members.aol.com/slo97/	Smith–Lemli–Opitz/RSH syndrome Information on clinical aspects	Family/professional
Smith Magenis syndrome		
http://www.kumc.edu/gec/support/smith-ma.html	Smith–Magenis syndrome Lots of information, links, references, support groups	Family/professional
http://www.bcm.tmc.edu/neurol/research/genes/genes10.html	Smith–Magenis syndrome Quick run-down of clinical features	Professional
http://www.cafamily.org.uk/Direct/s33.html	Smith–Magenis syndrome Contact a family site. Information and support	Family
Spinal muscular atrophy		
http://www.fsma.org/	Families of those with Spinal Muscular Atrophy Information, links, support	Family
http://www.fsma.org/booklet.htm	Families of Spinal Muscular Atrophy Quick guide to spinal muscular atrophy	Family/professional

Website	Description	Use
http://www4.ccf.org/health/health-info/docs/1300/1346.HTM	Spinal muscular atrophy User-friendly information	Family
Tuberous sclerosis		
http://www.tuberous-sclerosis.org/	Tuberous Sclerosis Association Latest news, information, links, funding	Family/professional
http://www.ntsa.org/guests/main.web	The National Tuberous Sclerosis Association Information, resources, services	Family
http://www.insteam.com/CT_TSA/TSdef.htm	What is tuberous sclerosis? People-friendly information	Family
Williams syndrome		
http://www.healthlinkusa.com/335.html	Healthlink USA, on Williams syndrome Links to several sites	Family/professional
http://www.williams-syndrome.org.uk/	Williams Syndrome Foundation UK Information, links, support groups	Family/professional
http://www.cafamily.org.uk/Direct/w15.html	Williams syndrome Contact a family site. Information and support	Family

Appendix II – Pathway to becoming a registered genetic counselor in the UK

In the UK, the majority of genetic counselors have a nursing background. The practice of nurses is governed by statutory regulation. However, as increasing numbers of counselors enter the field via a Master's program in genetic counseling, it is clear that a registration system is required. The registration system fulfills two purposes; the primary purpose is to protect the client by ensuring that practitioners have achieved a minimum standard of competence. Second, it ensures professional parity of genetic counselors, regardless of their pathway into genetic counseling.

The Association of Genetic Nurses and Counsellors (AGNC) has now established a formal registration process for genetic counselors. The registration system is based upon a set of competency standards. A code of ethics also exists for the guidance of practitioners.

There are two main routes of entry into the profession. Future applicants for registration as a genetic counselor must have either: (i) completed a recognized Master of Science degree in genetic counseling, which includes a substantial clinical and experiential component, and training in counseling skills or have a professional qualification in nursing or midwifery; or (ii) attained a relevant basic or postgraduate degree and a professional qualification (nursing, midwifery, social work), completed a minimum of 2 years' appropriate professional clinical experience in a health or social care setting and undertaken counseling training.

Assessment for registration will be based upon the applicant providing evidence that they are able to achieve the core competencies, following at least 2 years' experience in a genetic counseling role.

Current information on registration as a genetic counselor is available on the AGNC website (http://www.agnc.org.uk).

Appendix III – Credentialing system for genetic nurses in the USA

In the USA and Canada, genetic counseling is offered by both Master's graduates in genetic counseling and by genetics nurses. In general, genetic nurses would regard themselves as having a broad remit to provide health care (including genetic information) in a number of different healthcare settings, whereas genetic counselors may feel their single focus is on genetic counseling.

Genetic counselors are accredited by the American Board of Genetic Counselors.

The educational and experiential preparation of genetic nurses in the USA has now been formalized by a credentialing system developed by the International Society of Nurses in Genetics (ISONG). Following the statement issued in 1998 by ISONG and the American Nurses Association on the scope and standards of genetics clinical nursing practice, a portfolio based assessment system was devised.

Directions for constructing the portfolio can be obtained from ISONG. Briefly, the applicant must have:

1. A licence to practice as a registered nurse.

2. A minimum of 3 years' experience as a clinical genetic nurse at least 50% of the nurse's time spent in practice in genetics.

3. At least 300 hours of genetic practicum experience.

4. The nurse must submit:
- curriculum vitae;
- a log book of at least 50 cases;
- four case studies;
- peer review letters and performance verification;
- evidence of graduation from an accredited graduate nursing program.

The portfolio of evidence is assessed by the ISONG Credentialing Commission.

Detailed information can be obtained via the ISONG website (http://nursing.creighton.edu/isong/practice/credential.htm).

Appendix IV – Glossary of terms

Affected individual: A person who has the signs and symptoms of the genetic condition.

Allele: A copy of the gene at a particular locus. One allele is inherited from each parent.

Amniocentesis: Withdrawal of amniotic fluid from the amniotic sac, usually for the purpose of testing the fetal chromosomes.

Anencephaly: Failure of the anterior neural tube to close properly during very early intrauterine life resulting in the absence of the cerebral hemispheres and skull bone together with a rudimentary brain stem.

Aneuploidy: An alteration in the number of chromosomes, involving only one or several chromosomes rather than the entire set of chromosomes.

Anticipation: The phenomena whereby successive generations of a family manifest a genetic condition more seriously or at a younger age.

Assisted reproduction: Any artificial technique used to enable a pregnancy to be achieved (e.g. *in vitro* fertilization).

Autism: A form of mental disability characterized by failure to interact with others.

Autosomal dominant inheritance pattern: The inheritance pattern whereby one copy of a gene is mutated, this is sufficient to cause the disease to be manifested.

Autosomal recessive inheritance pattern: The inheritance pattern whereby both copies of the gene are mutated and the person develops the condition because they have no normal copy. Carriers of recessive conditions are usually unaffected.

Autosomes: The chromosomes that are present in equal numbers in both male and female of the species (in humans, the chromosomes 1 to 22).

Base pair: A pair of nucleotides, which are positioned opposite each other on the two strands of the DNA double helix. Adenine always pairs with thymine, and guanine with cytosine.

Carrier: A person who is generally not affected with the condition, but carries one faulty copy of a gene. Generally relates to heterozygotes in recessive or X-linked conditions.

Chiasma: The point where two homologous chromosomes cross over during meiosis.

Chorionic villus biopsy: Removal of cells from the chorionic villi (developing placental tissue).

Chromatid: One of two lengths of chromosomal material (sister chromatids) that are joined at the centromere during cell replication. Each becomes a new chromosome.

Chromosome: The physical structures into which the DNA is packaged within the nucleus of cells. The usual number of chromosomes in humans is 46.

Clinical genetics: The branch of the health service that is chiefly involved in diagnosis of genetic conditions and genetic counseling for families.

Codon: A triplet in the messenger RNA that provides the code for one amino acid.

Colonoscopy: Investigation wherein the rectum, sigmoid and large colon are viewed directly via an endoscope.

Consanguinity: The biological relationship between two individuals who have a common ancestor.

Consultand: The person seeking genetic information, not necessarily the affected person in the family, who is usually called the proband.

Cordocentesis: Removal of a sample of fetal blood from the umbilical cord during pregnancy.

Cytogenetics: The study of chromosomes, in the laboratory.

Delay – developmental or learning delay: A term used to describe the failure of the child to reach milestones in physical, mental, emotional or social development within the expected age limits.

Deletion: The omission of a part of the genetic material; the term can be used in relation to either a gene or a chromosome.

Diploid: Having two copies of each autosome.

Disjunction: The separation of the replicated copies of the chromosomes into two daughter cells during the second stage of meiosis.

DNA: Deoxyribonucleic acid. The biochemical substance which forms the genome. It carries in coded form the information that directs the growth, development and function of physical and biochemical systems. It is usually present within the cell as two strands with a double helix confirmation (see Chapter 4).

Dominant: See autosomal dominant.

Duplication: The abnormal repetition of a sequence of genetic material within a gene or chromosome.

Dysmorphic features: Physical features that are outside of the variability of the normal population. They may occur because of a change in the genetic code providing instructions for those features.

Eugenics: The study and practice of principles that will improve the genetic health and fitness of a population.

Exclusion test: A genetic test that uses samples from three generations of the family, to exclude the risk of a genetic disease or confirm a 50% or 25% risk.

Exon: A sequence of DNA that contributes to the protein product of a gene (see also intron).

Expansion: An abnormally large repetition of specific DNA sequences within a gene.

Expression: The way in which the gene mutation is manifested within an individual.

FISH (fluorescent *in situ* hybridization): A technique that uses both cytogenetics and molecular biology to identify subtle changes in chromosome structure.

Gamete: Cell formed in the reproductive organs from the germline, in humans either ovum or sperm.

G-banding: A technique of staining the chromosomes to enable identification by creating a different pattern of bands along each chromosome.

Gene: The fundamental physical and functional unit of heredity consisting of a sequence of DNA.

Gene therapy: Therapy that is based upon the principle of replacing or modifying a faulty gene with a normal copy, in the relevant tissues. The aim is to reduce or obliterate the effects of the genetic condition.

Genetic counselor: A person whose main professional role is to offer information and support to clients who are concerned about a condition which may have a genetic basis.

Genetic screening: This term usually refers to population screening for a genetic variation or mutation.

Guthrie test: Blood test performed in the neonatal period to detect infants at high risk of phenylketonuria. A test for congenital hypothyroidism is usually performed at the same time.

Hemoglobinopathy: A genetic condition that affects the structure of the hemoglobin molecule.

Haploid: Having one copy of each autosome.

Heterogeneous: Pertaining to more than one gene.

Heterozygous: Having two different alleles at a genetic locus, usually one normal and one faulty copy of a gene (see also homozygous).

Homologous pair: Two copies of the same chromosome.

Homozygous: Having two identical alleles at a genetic locus. In Mendelian diseases these may be copies of a gene that are either both normal or both faulty.

Hybridization: Attachment of one DNA sequence to an identical sequence. Hybridization is used to attach a DNA probe to a segment of genomic DNA.

Hydrocephalus: The presence of excessive cerebrospinal fluid in the ventricles of the brain, normally leading to an enlarged head.

Imprinting: The phenomena whereby the two copies of a gene have a different function, depending upon their parental origin.

Induced abortion: Termination of pregnancy.

Insertion: The introduction of additional genetic material into a gene or chromosome.

Intron: A sequence of DNA that does not contribute to the code for protein product, as the genetic sequence within introns is omitted when the mRNA is made.

Inversion: An alteration in the sequence of genes along a particular chromosome. In a paracentric inversion, the change occurs on only one side of the centromere. In a pericentric inversion, the centromere is involved and material will move from the long to short arm and vice versa.

Karyotype: A description of the chromosome structure of an individual (assessed during metaphase), including the number of chromosomes and any variation from the normal pattern.

Linkage: The phenomenon whereby alleles that are physically close together on a chromosome will tend to be inherited together. This allows for a technique of genetic testing that tracks a specific copy of a gene through a family.

Locus: The position of a gene, a genetic marker or a DNA marker on a chromosome.

Maternal serum screening: A method of detecting a relative risk of Down syndrome in a pregnancy using biochemical testing of the mother's blood.

Meiosis: The production of gametes (haploid cells).

Mendelian disorder or **Mendelian condition:** A genetic disorder caused by a single gene fault, following a dominant, recessive or X-linked pattern of inheritance.

Messenger RNA: The sequence of base pairs that transfer the genetic code from the DNA to a functional protein.

Microdeletion: A minute deletion of chromosomal material that is usually not detectable down the microscope and has to be identified using a method such as FISH.

Mismatch repair gene: A gene whose function is to detect and repair errors in DNA transcription.

Mitochondrial DNA: The genetic material in the mitochondria, outside the nucleus of the cell.

Mitosis: The production of somatic diploid cells.

Molecular genetics: The study of genetic material at a molecular level, including DNA studies.

Monosomy: Having only one copy of a particular chromosome.

Mosaicism: Having more than one cell lines with different chromosomes or expressing different genes.

Multifactorial: A condition is said to be multifactorial if both genetic and environmental influences are thought to be causative.

Mutation: A gene sequence variation that is found in less than 1% of the population. The mutation may cause a change in the protein product of the gene, and therefore cause health problems for the person concerned.

Neonatal death: The death of a baby who has shown signs of life, before the age of 28 days.

Neural tube defect: An abnormality of the spinal column or cranium (spina bifida or anencephaly).

New genetics: A term used to denote a change in the field of genetics, where the focus shifts from rare conditions caused by a fault in a single gene (including the 'Mendelian' conditions) to application of genetics to common diseases.

Non-directiveness: A model of counselling used in genetic counselling, which emphasizes the right of clients to make decisions without coercion from others.

Non-disjunction: Failure of the two copies of chromosomes to separate effectively into the two daughter cells.

PCR or **Polymerase chain reaction:** A laboratory method of manufacturing many copies of a sequence of DNA.

Pedigree: Family tree.

Penetrance: The extent to which specific gene mutations are manifested within an individual.

Pharmacogenetics: The science of using the genetic variability in the population to target medication more effectively.

Polygenic: Relating to a number of different genes, e.g. a disorder is polygenic if it could be caused by a combination of mutations in several different genes.

Polymerase chain reaction: *see* PCR.

Polymorphism: Normal variation in sequence of DNA in a gene, differs from mutation in that it is usually found in more than 1% of the population.

Polyp: A small tumor growing from the surface of mucous membrane.

Population screening: Using a test to assess the risk or presence of a genetic disease in an entire section of the population e.g. neonatal screening for hypothyroidism.

Proband: The affected person in the family or the person who is seeking genetic advice.

Probe: A sequence of manufactured DNA that attaches to an identical sequence in the genomic DNA for the purposes of genetic testing or research.

Recessive: See autosomal recessive.

Reciprocal translocation: Exchange of chromosomal material between at least two chromosomes.

Recombination: The creation during meiosis of a new chromosome or sequence of DNA, which is a unique combination of the parent's maternal and paternal DNA.

Recurrence risk: The chance that a genetic condition will occur again in offspring or siblings of an affected person.

Restriction fragment length polymorphisms (RFLP): Fragments of DNA within a gene that have a normal variability in size when cut with specific enzymes.

Robertsonian translocation: An attachment of two acrocentric chromosomes end to end at the centromere.

Scanning: Investigation of physical structures using ultrasound device (sound waves).

Single nucleotide polymorphisms (SNPs): Polymorphic markers that detect single base changes in the DNA sequence.

Somatic: Relating to cells other than the germline.

Southern blotting: A laboratory method of DNA analysis.

Spina bifida: An interruption to the spinal column, with possible herniation of the spinal cord and meninges (myelomeningocoele). One form of neural tube defect (another being anencephaly).

Spontaneous abortion: Loss of a pregnancy without interference, miscarriage.

Stillbirth: A fetus of more than 24 weeks gestation who is born dead.

Syndrome: A number of physical features or abnormalities that fit a recognized pattern.

Teratogen: A substance that may harm the developing fetus.

Translocation: An alteration in the usual structure of a chromosome, wherein part or all of one chromosome is attached to another.

Trinucleotide repeat: A sequence of three bases that is repeated more than once at a site within a particular gene.

Trisomy: Having three copies of a particular chromosome.

Tumor suppressor gene: A gene whose normal function is to prevent the overgrowth or abnormal growth of cells.

Uniparental disomy: Inheritance of both copies of a particular chromosome from one parent only.

VNTR polymorphisms (variable number tandem repeats): Variations in the number of repeat sequences of DNA at a specific locus.

X-inactivation: In human females the early random inactivation of one of each of the X chromosomes, allowing expression of genes only on the active X chromosome.

X-linked inheritance pattern: A pattern of inheritance whereby the mutated gene is on the X chromosome, of which males have one copy and females have two.

Appendix V – Answers to test questions

Chapter 1

Answer

Q1. Genetic counselling interactions will be tailored to the clinical situation and needs of the client. In general the following components should be considered:
- diagnosis;
- information giving;
- evaluation of risks;
- presentation of options;
- ongoing support.

Chapter 2

Answers

Q1. Using Bayes theorem 1/41 or just over 2%

	Affected	Not affected
Prior probability	1/4	3/4
Conditional probability	10% = 1/10	1
Joint probability	1/10 × 1/4 = 1/40	1 × 3/4 = 3/4 = 30/40
Final probability	1/41	30/41

Q2. Using Bayes theorem 2/7 or approximately 29%

	Affected	Not affected
Prior probability	1/2	1/2
Conditional probability	40% = 2/5	1
Joint probability	1/2 × 2/5 = 2/10	1 × 1/2 = 1/2 = 5/10
Final probability	2/7	5/7

Q3. Spinal muscular atrophy is recessive. To calculate carrier frequencies use the Hardy–Weinberg equation

$q^2 = 1/14\,400$

q = square route of $1/14\,400 = 1/120$

$2q = 2/120 = 1/60$

The chance that someone in a family with no history of this condition will carry the gene is $1/60$

The risk to a couple of an affected child if one carries the gene and the other partner is at population risk is

$1 \times 1/60 \times 1/4 = 1/240$ approximately 0.4%

Points to consider.

- If the prior probability of being affected is high and a negative test does not rule out the possibility of having the gene (does not have a high negative predictive value) as in example 2, a negative test result will not reduce the risk substantially. Whereas in example 1, because the prior probability was lower and the negative predictive value of the test was higher the final risk of Ben having polyposis was lowered by the result of the colonoscopy.
- With recessive conditions unless the gene frequency in the population is high the risk to the offspring of a carrier is low provided their partner has no history of the condition in their family.

Q4. Points to consider in your reflection.

- Family trees are sensitive confidential information.
- Each individual in a family may hold different information at different levels of knowledge.
- Families may not have accurate information about a diagnosis in other members.

Chapter 4

Answers

Q1. An increase in non-disjunction during meiosis the mechanism for which is not known.

Q2. (a) Turner syndrome.
(b) Edward syndrome.
(c) Down syndrome.

Q3. (a) 46,XX
(b) 47,XXY
(c) 47,XX, +13

Q4. DNA is the molecule which forms the genetic code and allows replication and transmission of the code through cell division. This code is in the form of genes which code for specific proteins.

Q5. One explanation for this finding is non-paternity. If this was discovered as part of a research project it would be unlikely to be disclosed. If it was discovered during clinical testing the situation would require very sensitive and careful handling. What is actually disclosed would depend upon the exact clinical situation. Consider as a practitioner what action you might take in such a situation.

Chapter 5

Answer

- Taking folic acid to reduce the risk of neural tube defects.
- Is there is any significant medical history in the mother that needs to be considered?
- Is the mother on any medication that may be teratogenic in which case specialist advice should be sought e.g. anti-convulsants, acne preparations containing vitamin A?
- Is there any significant family history e.g. learning difficulties, muscular dystrophy, any other genetic condition that might need referral to clinical genetics?
- Is the mother older? In which case she may wish to prepare for making a decision about prenatal diagnosis.
- General health advice regarding diet, smoking, alcohol, recreational drug use.

Chapter 6

Answer

Points to consider:

- People in this sort of stressful situation may only hear part of what is told to them. The first thing they are told may be what they remember.
- Although any decision a couple make is individual to them it is important that all the information is accurate, up to date and presented in a non-directive way.
- Language is very powerful, the use of words such as monster, abnormality and syndrome may have many different meanings and should be used carefully.
- It should be understood that the purpose of any prenatal test is to detect abnormality not to provide reassurance about normality.

Chapter 7

Answers

Q1. Population screening is the process of going out to a population and offering a test to indicate who in that population might be at risk of developing a specific condition. Diagnosis of the condition usually requires further confirmatory tests. Screening should only be undertaken where a condition is an important health problem, where there is opportunity to intervene early and initiate effective therapy and where the screening test and subsequent diagnostic tests and treatment are acceptable. Genetic testing is where a test is used in an individual where a diagnosis is already being considered and where the individual themselves has expressed concern.

Q2. (a) karyotyping fluorescent *in situ* hybridization (FISH)

(b) Check the parents' chromosomes. If they are normal the risk is low. If one of the parents has the deletion the risk of passing the deletion on is 50%. The actual problems associated with the deletion are more difficult to predict.

Q3. A malformation is a primary alteration in development i.e. the organ or tissue is programmed to develop abnormally, a deformation is where tissue or an organ is programmed to develop normally but external forces e.g. lack of amniotic fluid cause abnormal development.

Chapter 8
Answer

Points to consider
- There is no right or wrong answer in this situation.
- Clear discussion of the conflicting rights can sometimes clarify things.
- Joe cannot be coerced into giving a blood sample.
- Neither Marie's nor Joe's rights take precedence.
- If Joe was a younger child the purpose of testing would probably be to make a diagnosis in him. This may be sufficient reason to proceed with the blood test, since it may lead to changes in management or treatment which would be to his personal benefit.

Chapter 9
Answers

Q1. Points to consider
- There are two issues here: the woman's decision not to have a colonoscopy herself and her unwillingness to allow her family to be informed.
- The primary duty of the practice nurse is to her patient, protecting her confidentiality. However in line with professional codes of conduct confidentiality can be breached if the potential for harm to another is serious. This would require careful discussion with colleagues, the patient of one's professional body.
- In practice exploration of the reasons for the woman's decision both for herself and her siblings may resolve the situation.

Q2. Although haemochromatosis is a recessive condition, the carrier frequency of 1/10 is high means that the child of somebody with haemochromatosis has a 1/20 chance of being homozygous. The penetrance in homozygous individuals is less than 100%, but the fact that there is effective treatment which prevents the serious consequences of the disorder means that first degree relatives should be offered testing.

Q3. Points to consider

- The cancers in *BRCA1* families are adult onset therefore the risk to a child of developing serious disease during childhood is no higher than any other child.
- The individual's experience of cancer in their family will color how they perceive the risk.
- The knowledge that one may pass on a serious disease causing gene leads to powerful emotions of guilt and blame.

Index